Critical Issues Facing America

Fred McKinney

Critical Issues Facing America

Selected Essays from a Black Economist, 2016–2025

palgrave
macmillan

Fred McKinney
Cofounder
BJM Solutions, LLC
Upper Marlboro, MD, USA

ISBN 978-3-031-93216-8 ISBN 978-3-031-93217-5 (eBook)
https://doi.org/10.1007/978-3-031-93217-5

This Palgrave Macmillan imprint is published by the registered company Springer Nature Switzerland AG
The registered company address is: Gewerbestrasse 11, 6330 Cham, Switzerland

Foreword

I am pleased to recommend Dr. Frederick W. McKinney's commentaries on dozens of social and public policy issues for your reading pleasure and instruction. Dr. Fred, as many of his business associates and friends refer to him, embodies a superb combination of personal talents, training, and experiences that make this book a rare learning experience for readers. Let me introduce the author of this cavalcade of essays brimming with wit and social policy wisdoms to you.

I first met Fred McKinney on the street outside the economics department during the first few days following my arrival at Yale. Fred was literally packing books into the trunk of his car to depart for Washington, D.C., where he would spend a year working on the research staff of the President's Council of Economic Advisors. Because we would not actually begin to know one another until a year later when he returned to full-time duties as a graduate student in the economics department, that first meeting was a brief encounter. It was not indicative of the important future friendship that would enrich both of our professional and personal lives. Upon his return, we bonded in two important roles; professionally, I was advisor to his doctoral thesis and recreationally on the basketball court where the economics team terrorized opponents (I was good, Fred was the best player on campus including the Yale varsity). I mention this aspect of his background because Fred's blend of nerdiness and athleticism are illustrative of the unique array of talents and experiences that inform his commentaries. A blending of academic competence and practical world skills informs each of his essays. When Fred left Yale, doctorate in hand, he had been trained to become a professional economist and that academic training was both tempered and strengthened by the deep practical experience he had gained at the Council of Economic Advisors where he participated and observed first-hand the inside workings of economic policy in the inner sanctum of the nation's governmental policy center. His initial professional employment as an Assistant Professor at Brandeis University's Heller School of Public Policy allowed him to hone these skills to an even sharper edge which shines through in each of the chapters of this book. But Fred's technical training as an economist at the highest levels of academia and his practical experience in policy-making are only two attributes of his that make this book a gem of insights.

I referred earlier to Frederick as a nerd and the exemplar of real-world practices. Even as a graduate student, he exhibited entrepreneurial talents launching business ventures that took him to the vast African continent. Post-graduation becoming a business owner, he gained the skills needed to market his product to buyers and to meet the demands of making a payroll, skills that can only be learned by doing. Later, as a consultant to government organizations ranging from small towns to large cities, from states to the federal government, and to private businesses ranging in revenue from six figures to many billions, he developed further skill sets that enabled him to be the successful director of a nationwide organization creating contracting opportunities for minority and women businesses with large firms including Fortune 500 companies. Frederick W. McKinney's range of experiences is incredible. Hence, for just three examples, when he discusses the condition of Black and minority businesses and Black wealth in Chapter 2, analyzes economic markets from a macroeconomic lens in Chapter 3, or addresses Black Americans' political situation and their possible strategies and prospects going forward in Chapter 8, we can be sure his insights are not just academic theories that may or may not have practical utility. These commentaries spring from a rare fountain of knowledge and experience founded on superb academic training and practical business and policy experience. Each chapter offers valuable insights that are thought provoking and useful whether one agrees with its conclusions or not.

Dr. Gerald D. Jaynes, Ph.D.
A. Whitney Griswold Professor
of Economics
Black Studies, and Urban Studies
Yale University
New Haven, CT, USA

Preface

The essays collected *in Critical Issues Facing America: Selected Essays from a Black Economist, 2016–2025* represent nearly a decade of reflection, analysis, and conviction. They are born out of my lifelong commitment to examining the forces that shape American society—economics, politics, race, and justice—and to sharing those insights in ways that are both rigorous and accessible.

This project would not have been possible without the people who walked alongside me on the journey. I am deeply grateful to my editors at Hearst Newspapers, Hugh Bailey, and John Breunig, whose sharp editorial judgment, patience, and encouragement made these essays stronger and more focused. Their guidance reminded me that clarity and accessibility are as important as substance.

I also thank my new friends at Palgrave Macmillan Publishing for having the confidence in my work as original and helpful to the public discussion taking place in America and around the world. I particularly want to thank Bronwyn Geyer at Palgrave Macmillan for her encouragement and support.

To my wife, Ivy Thomas McKinney, I owe a profound debt of gratitude. She has been a constant sounding board, an honest critic, and a source of unwavering support. More than once, she reined in my worst impulses and reminded me of the bigger picture. Her love and wisdom sustained me through the long hours of writing and reflection.

I also thank my father Rufus McKinney, Sr. who this year celebrated his 95th birthday and his wife, my second mother, Glendonia McKinney. They provided a rigorous intellectual environment while maintaining the warmest most welcoming home anyone could dream of. My father, an accomplished author himself, provided thoughtful suggestions and editorial advice. He read the entire manuscript, and we had some interesting and sometimes heated discussions on much of the material included. His intense interest confirmed for me that this book may resonate with readers, whether they agree with me or not.

I also recognize my siblings; Rufus McKinney, Jr., Anne Brobby, and Paula Jordan. Special thanks to my children Warren McKinney and his wife Sahar Lofti-Emran, and Dana McKinney White, her husband David White, Jr. While my grandchildren maybe too young to provide editorial advice, they made up for it by providing inspiration. Thank you Aydin McKinney (7), Kian McKinney (5), and David E. White, III (Emi) (15 months). My nieces, nephews, cousins, and in-laws are too numerous to name, but they also contributed even if only by reading sections of the book. I love them all.

I also wish to honor my mentor, colleague, professor, business partner, and friend, Dr. Gerald Jaynes. His example has shaped not only my scholarship but my life. Gerald has shown me how to marry intellectual rigor with moral clarity, and how to pursue truth while never losing sight of humanity. I am privileged to have learned from him and to have worked alongside him.

I also want to thank my ancestors; the ones I know and whose memory is forever imprinted in my being as well as the ancestors whose names are lost to time and the legacy of slavery. I want to lift up my grandparents George McKinney, Sr. Mama McKinney (she had 13 surviving children), Mattie Christine Olma Katie Lee Brown Davis Colclough, and Silas Edward Davis. My grandparents were guiding lights in my life as a child when I visited them in Pine Bluff, Arkansas. My Aunts Esther, Ruth, Gail, Lola, Izora, and Jo Marva (who was named for Joe Louis and his wife Marva because she was born on the day Joe Louis defeated Max Schmeling) were all loving and caring professional Black women when it was not easy to be either a Black woman or a Black woman professional.

I do not remember much about my maternal great grandfather, Papa Brown, but I know his story of being born in Mississippi in 1863, the son of his owner. His mother, my great-great grandmother was owned by his White father (my great-great grandfather) who had the audacity to name my great grandfather, Jefferson Davis Brown after the first president of the Confederate States of America. Papa Brown told the story of killing his White half-brother for the rape of his sister. He left Mississippi with the help of his White father. He told the story in Biblical fashion. According to Papa Brown, he "parted the waters" meaning he crossed the Mississippi River into Arkansas to escape the men who were chasing him and would have certainly lynched him for his action.

I reserve the highest honor to my mother, Dorothy Ann Davis McKinney Wright who from a very early age would ask me what I thought about current events that she would read to us from the morning paper. I am the product of all these ancestors, and I am eternally grateful for their love and guidance.

This book is, in every sense, a labor of love. It is a love for ideas, for dialogue, for the pursuit of equity, and for the America that we can become. My hope is that these essays spark conversation, challenge assumptions, and contribute, in some small way, to a better understanding of the complex issues before us.

Upper Marlboro, MD, USA Fred McKinney

Contents

CHAPTER 1

Critical Issues: Economics and Race

Abstract Critical Issues Facing America: Selected Essays from a Black Economist, 2016–2025, provides an expansive review of political and economic topics from the author, Frederick McKinney, a Ph.D. Yale University trained economist. The book covers these topics from the perspective of a Black economist. The book is designed as a supplement to economics courses, but should be of interest to the general reader interested in public policy and economics.

Keywords African-American · Black · Economics · Race · American History · Policy · Discrimination · America

The critical issues facing America have changed with changes in technology and population, but fundamentally, the problems this book addresses would be familiar to any economist or researcher who looked at America's issues fifty years ago. A short list of those problems would include unemployment, poverty, hunger and food insecurity, homelessness, housing quality and cost, healthcare access, quality, and costs, transportation, discrimination and segregation, access to technology, privacy, government policy, the environment, international relations, political tension, crime, and immigration. This book contains short essays and editorials on many of these issues. At the center of all these issues,

F. McKinney, *Critical Issues Facing America*,
https://doi.org/10.1007/978-3-031-93217-5_1

however, is the distribution of income and wealth and how that distribution impacts the lives of ordinary Americans. The distribution of wealth and income interacts with political forces that determine economic policy. Even before Supreme Court cases like Citizen's United that granted corporations rights previously reserved to natural humans, corporate power and elites have disproportionately impacted the rules of the game to favor their interests. Fifty years ago, that power operated behind the scenes to keep Americans convinced that their success in life was the result of their own decisions and luck. And that their failures were not the result of institutions designed for the continued success of those at the pinnacle of the pyramid. Now the wealthy have come out from their dark places and have asserted their rights to blatantly turn American democracy into an oligarchy and end the charade.

Addressing the long list of critical issues facing Americans cannot be accomplished without addressing the institutions that create and support the interests of the wealthy. I firmly believe relying on marginal changes in policy under the current institutions (the American pyramid) to radically change the conditions of low-income Americans, particularly Black Americans, is asking for more of the same gradual progress that is captured by one stark economic fact: too many Americans are poor and struggling.

As a Black economist, my focus and concerns are for Black Americans. It is important to note at the outset, that problems Black Americans face are largely the same as the problems other Americans face, particularly those near the bottom of the American pyramid. My focus on Black Americans is largely because of my familiarity to this community, but I encourage others to read my words in the context of these problems in their own communities. There is similarity in conditions across communities even if there are differences in what are thought to be the sources of similar problems. Because of the history and legacy of slavery however, Black Americans have a unique place in American society that is not shared by any other racial or ethnic group. This uniqueness is addressed throughout the book.

There is nothing new about economists writing about social policy and human behavior. The modern science of economics began in 1776 with the publication of Adam Smith's classic book, *An Inquiry into the Nature and Causes of the Wealth of Nations.* I purposefully use the full title of the book instead of its shortened more popular version, *The Wealth of Nations*, because the full title describes the scientific principles of observation, measurement, hypothesis formation, testing, conclusions, and implications for public and private policy, that distinguish science from pseudo-science. Smith's concern about the general level of economic well-being and how that is related to the organization of scarce resources, ushered in the study of economics. Smith's book covered a range of topics that are still core components of economic science and knowledge today. The theory of the capitalist's firm, the organization of industries, the behavior of companies based on the structure of industries they operate in, the advantages of specialization, the role of the entrepreneur, the behavior of capitalists, the source of profits, the contribution of workers, the effect of international trade, agricultural policy, the role of government, the role of technology and innovation, demand, supply, and market equilibrium, and the transformation of society are all topics covered in this seminal book.

If there is a second book that has influenced my thinking and general economic thought that book would be John Maynard Keynes, *The General Theory of Employment, Interest, and Money* published in 1936. The General Theory was written during the height of the Great Depression. Economic and political stress forced economists to reevaluate their models of how society works and how institutions should respond. Keynes showed that during "normal" times, markets efficiently supply consumer's goods and services they desire and create employment and financial conditions consistent with social harmony. However, capitalist economies periodically deviate from their ideals and instead of Smith's invisible hand leading to the best outcomes, society can devolve into an equilibrium that is not desirable by capitalists, workers, consumers, or government. Keynes' solution at the time was economic blasphemy. His critics, the classical economists believed that allowing wages to fall to find a new equilibrium was the appropriate policy for eliminating unemployment. Keynes argued the government should borrow money to spend on bringing about more employment. In his view, waiting for wages to decline to exert their equilibrating effect on unemployment made no sense. The problem was inadequate aggregate demand. Under these circumstances, no reduction in wages would be sufficient to induce companies to hire unemployed workers.

Keynes also had similar disdain for relying on lower interest rates to increase investment by firms during periods of economic crisis. Keynes argued capitalists were like lemmings who are risk adverse actors and are even more so during an economic crisis, and lowering interest rates during an economic crisis will not convince them to invest in new plants, equipment, and technology. Again, inadequate aggregate demand is the source of the problem of underinvestment not interest rates that are too high.

Keynes was describing a market failure that was not self-correcting and required government intervention. To Keynes, there was nothing "natural" about market systems that led to social harmony. In fact, Keynes and Smith had something in common when it came to the source of market failure which was monopoly power. Smith said:

> People of the same trade seldom meet together, even for merriment and diversion, but the conversation ends in a conspiracy against the public, or in some contrivance to raise prices.

Keynes stated in his concluding chapter:

> The outstanding faults of the economic society in which we live are its failure to provide for full employment and its arbitrary and inequitable distribution of wealth and income.

The Smith branch of classical and neoclassical economics relies on the correcting forces of markets and the unintended consequences of interfering with market forces, has once again emerged as the dominant view in Western economies including the United States. This market focused approach to solving problems associated with income wealth inequality, stable prices, economic growth, public education and welfare, environmental challenges, racial and gender inequalities, tax and spending policy, and military defense are fundamentally and irreconcilably different from the economic philosophy that believes markets sometimes fail and when they do, economic and social benefits are substantially reduced for the majority of the population.

In my own training as an economist, I started at U.C.L.A. which in the 1970s was squarely in the Chicago School of economics, then led by Nobel Laureate Milton Friedman. As an undergraduate, I got a strong dose of market based economic analysis. One of the major contributions of the Chicago School approach was to distinguish between positive economics and normative economics. Positive economics is the study of what will (or is likely) to happen when consumers and firms face changing prices or other market circumstances. Normative economics focuses on what should happen. The key words that distinguish positive from normative economics are "should" and "will." Any observer of the human condition may have an opinion of what should happen. Those opinions may or may not be consistent with economic analysis that focuses on what will happen.

Then as a graduate student at Yale University, I received a strong dose of Keynesian economic analysis. One of the leading American proponents of Keynes, James Tobin, also was awarded the Nobel Prize in economics and was personally instrumental in me coming to study at Yale. I found the Keynesian approach much more in keeping with my own biases towards intervention as opposed to waiting for market forces to correct social problems, particularly problems of income and wealth inequality.

Much of this book represents my views on what should happen, based on what will happen if nothing changes. My focus is on the effects of economic and social policy changes on all Americans, but I am laser focused on the impact of policies on Black Americans. In many of the articles comprising this book, the audience was multi-racial. I would often use the term "minority" to include Blacks, Hispanics, Asians, Aboriginals. I refuse to include women within my definition of minority, based on factual considerations and clarity. Some women are also minority, but not all women are. While women have a unique set of problems, the problems facing Black Americans will not be solved if the determining characteristic is gender. White women's problems given their relationship to being the daughters, wives, and sisters of the economically dominant group of White men cannot be the basis for analyzing the problems facing the Descendants of Stolen People (DSP). These are relationships that are not common among Black Americans.

Even the narrower inclusive definition that includes other non-White racial and ethnic people can result in policies designed to support "minority" communities that are not maximizing the economic and political interests of Black Americans. Hispanic Americans is also a problematic term when it comes to understanding the role of race and ethnicity in America. This catch-all term includes Cubans, Mexicans, Central Americans, South Americans, and Europeans from Spain and Portugal. The term also includes people who are Black, White, and Aboriginal. The term is so broad that it lacks an ability to target policies to people with specific conditions.

There are similar problems with the category of Asian Americans. These include peoples from China, Japan, India, South Asia, and the Pacific Islands. In statistical parlance, the variation within this group (and the same can be said for Hispanics) is greater than the variance across groups.

Definitions of race and ethnicity are a much different dynamic for White Americans. All White Americans are descendants of European immigrants. They came from Britain, Ireland, Germany, Scandinavia, Poland, Russia, Italy, Spain, Portugal, and elsewhere in Europe. They all had the ability to racially assimilate into American society as members of the dominant White racial group. This was not an option for most Descendants of Stolen People (i.e., Black Americans).

This book is not a compilation of the conditions facing Black Americans. But I do think it is important to provide the reader with a theoretical model of how I see Black Americans in American society. This model will explain the conclusions I reach in the broad subject matter that I cover in the book and explains my thoughts about what Black Americans should be doing now and in the future to enhance the collective wealth of the Black community.

1.1 A Model of Black American Economic and Political Mobility

One of the main features of both the Smithian and Keynesian schools of thought is the focus on the individual. Traditional Smithian economics, also called Classical and its Neo-Classical successors look at society as a combination of millions of self-centered economic agents seeking to maximize their incomes, their happiness, their wealth, or their profits. The firm, the consumer, the worker, and the entrepreneur are the key building blocks of the capitalist economy. To understand how capitalist markets work, you need to understand consumer behavior. Economic science assumes consumers are utility maximizers, each attempting to get as much happiness for themselves as possible. (The model does allow for consumers to include the happiness of others, but for the most part, the focus is on their narrower personal interests.) The theory of the firm assumes that capitalists are profit/wealth maximizers, each trying to generate as much profit as possible. The theory of the worker assumes that workers are utility maximizers who tradeoff leisure time for goods and services.

My model starts not with the individual, but with the collective society. America is not the simply to sum of its parts. America is first and foremost a hierarchical economy built on a base of immigrant labor and the labor of stolen people, stolen land from Aboriginal peoples, the accumulation and concentration of financial capital aided by political and military power. The American economy (and in many respects the world economy) is pyramidal in shape with a few elites at the top and the masses of people at the bottom. Social and economic mobility is limited but common enough to inspire acquiescence by the masses to the legal and political norms that keep society from exploding into revolution and class and or racial warfare.

The original American pyramid was influenced by the British class system and aristocracy. The 1776 American revolution brought together the American elite and White farmers who saw the economic and political opportunity independence offered to them and their progeny. Americans did not need the British aristocracy when they had their own. Most of America's founding fathers were wealthy landowners. Additionally, twelve of the first eighteen president (two-thirds) owned slaves, including Washington, Jefferson, Madison, Monroe, Jackson, Taylor, Van Buren, Harrison Tyler, Polk, Andrew Johnson, and Grant.

In this imported system of a landed elite, slavery, stolen land and the labor of stolen people was at the center of wealth creation. Stolen Africans were at the bottom of the American pyramid in 1776 and continue to be at or near the bottom of American society today.

But like all hierarchical systems, there was movement up the pyramid (as well as down) that was supported by the mythology of Adam Smith's market economy. In that mythology, if you worked hard, saved your money, improved your skills, stayed out of legal and moral trouble, in a generation or two, you too could move from the bottom to the top, or as rapper the Drake sings; "started at the bottom now I'm here." And if you did not make it to the top, you could at least put some social distance between yourself and those at the bottom.

The American racial caste system prohibited that movement for Blacks, or as I like to call Black Americans, the Descendants of Stolen People (DSP) during slavery, when it was illegal for slaves to own property, serve on juries, pray to their African Gods, read books, or simply control how they spent their time. Legal slavery started before the founding of America in 1619 and lasted 246 years until 1865. And after the end of slavery, except for a brief 10-year period of Reconstruction, the era of American apartheid, commonly known as Jim Crow lasted between 1877 and 1964/65. The end of this period was marked by the passage of the Civil Rights and Voting Rights bills. For all the 406 years, Europeans have lived on what is now American soil, the Descendants of Stolen People have served as legal slaves or the victims of Jim Crow America for 85 percent of that time. And for the other 15 percent of that time, only during the ten years of Reconstruction, can you unequivocally state that DSPs experienced some semblance of American political equality guaranteed by the founding fathers. The 60 years since the passage of the Civil Rights Act and the Voting Rights Act can best be characterized as a struggle between the aspirations of Black Americans and (some) White Americans who view those Black aspirations as coming at their expense. Eliminating the most recent 60 years as a period of mixed results, the entire American experience can be viewed as 396 years out of 406 years of institutional oppression or constant White backlash.

In this model of society, individuals move up the pyramid and some move down. The American Dream is for hard-working Americans to have greater incomes and wealth than their fathers, mothers, and grandparents. This dream was possible for White immigrants, many of them did not speak English. In a generation or two, they became proficient in English and in American culture which allowed them to assimilate or at least code shift from their immigrant roots to becoming full-fledged White Americans, indistinguishable from other White Americans whose origin could not and cannot be established by simply looking at them. No matter how well Black Americans spoke English or adopted the ways of the dominant culture, full assimilation was and is, impossible.

The pyramidal structure of American society in the context of racial caste supported White supremacists' views that Black Americans were inferior. This belief system was also instrumental to American elites who could garner the support of White workers who were told that no matter how poor and desperate they were, they were better off than the most educated, cultured, wealthy Descendant of Stolen People. The White elite supported this view by institutionalizing work where Blacks rarely supervised Whites or had power over White workers. The American pyramid was Blacks at the bottom and White on top of them. Blacks were allowed to move up, but only within their racial group. The election of Barack Obama in 2008 was a shock to the American pyramid and created the Trump backlash. Trump's wanting to Make America Great Again, is simply an attempt to go back to when all White Americans were on top, and all Black Americans were at the bottom. Diversity, Equity, and Inclusion (DEI) are clearly inconsistent with America's racial pyramid if talented Blacks, given the opportunity to succeed, do so at the expense of mediocre Whites.

In the final chapter, I will cover what this model of America means for how Black Americans should respond to Trump and the current economic and political situation. I want to say at the outset that it is my firm belief that the racial animus between Blacks and Whites will never be resolved without Black Americans (DSP) establishing a strong economy that serves the needs of Black Americans without the dependence on the White dominated federal government, large White owned enterprises and banks, and White small- and medium-sized businesses. It is not a realistic goal for Black-owned businesses to immediately replace the large global companies that dominate the American pyramid. The goal is for millions of Black-owned enterprises to replace the small- and medium-sized White owned and other ethnic businesses that dominate our community.

1.2 Black Social Theory

Anyone interested in understanding and effectively addressing the problems facing DSP (Black Americans) must have a firm understanding of Black Social Theory (BST). BST comprises the thoughts and movements of DSP since the beginning of the Black experience in America. BST is essentially the same as the more popular term, Critical Race Theory (CRT). Both look at social outcomes as a function of institutions and the effect these institutions have on human conditions. The reason CRT has been such a lightning rod in conservative circles is that conservatives like to think it is the action of individuals not institutions that determine social outcomes. To conservatives, you work hard, you are born "superior," maybe you have a little luck that determine your income, your wealth your position in the American pyramid, not the pyramid itself and the institutions that created and maintained that pyramid. For conservatives, blaming institutions takes away from their self-image of creating their own success, and that anyone who has not experienced the benefits of modern capitalist society have only themselves to blame.

Thought leaders also include action leaders who may not have recorded their words in books or in the halls of academia. These would include men and women whose names, actions, and beliefs rarely make it into economic literature or textbooks. These include Prince Hall who started the Prince Hall Freemasonry in 1784 to Ta-Nehisi Coats. In between these thought leaders are luminaries like Nat Turner, Denmark Vesey, Wiliam Monroe Trotter, WEB Dubois, Marcus Garvey, Elijah Muhammad, Ida B. Wells, Martin Luther King, Malcolm X, Angela Davis, Maya Angelou, H. Rapp Brown, Fred Hampton, Eldridge Cleaver, Huey Newton, Carmichael, Bob Marley, Isabel Wilkerson, Nicole Hannah-Jones, Eddie Glaude, Ibram X. Kenji, and others. These men and women's life work centered on how to improve the conditions of DSP. They did their work largely by documenting the history and stories of DSP. They may not have been trained as economists, but their work clearly demonstrated that they understood the importance of economics and social policy and the unique position DSP have in the American pyramid.

My work as an economist has been influenced by these thought leaders as much if not more than the work of formal economists. A synthesis of economics and BST may be one of the lasting contributions of this book for future writers. It is my firm belief that the next generation of BST leaders will more formally incorporate BST and traditional economics. But one thing that is clear from my decades of experience in higher education is that this synthesis will not originate in departments of economics. This synthesis must come from within the Black intellectual community.

1.3 Who Should Read This Book

This book is written for a general audience. My hope is that economic faculty might use the book as a supplement to the standard textbooks so that students might be exposed to some different thoughts when it comes to economic and social policy. A second audience is the Black reader. Movements of social change need literature produced by members of the community seeking change. A third audience is the economic policy community. These leaders often have limited exposure to strains of thought outside of their comfort zone. John Maynard Keynes famously said, "Practical men, who believe themselves to be quite exempt from any intellectual influence, are usually the slave of some defunct economist. Madmen in authority, who hear voices in the air, are distilling their frenzy from some academic scribbler of a few years back."

1.4 Organization of the Book

This book is a compilation of commentaries published in a number of different newspapers from 2016 to 2025. A wide range of topics was covered over this period, but there were some common themes. The chapters that follow are organized loosely around those common subjects, although there is the inevitable overlap and there are some topics that I felt deserving to be included that did not fit neatly into one of the chapters.

Chapter 2 focuses on Black wealth and the plight of Black and minority businesses. This focus on MBEs is largely due to the work that I was doing at the Greater New England Minority Supplier Development Council (GNEMSDC), the regional affiliate of the National Minority Supplier Development Council (NMSDC). After becoming CEO of the Connecticut Minority Supplier Development Council in 2001, I started a weekly newsletter. I ended each newsletter with a short message, I called "And Finally." Over the period 2001 to 2015, I published four compilations of those commentaries in books titled And Finally through And Finally 4.

My work with MBEs and corporate supplier diversity continued at the Tuck School of Business at Dartmouth where I served as a Managing Director for Executive Education. My work included developing and delivering non-degree training programs for MBEs and corporate supplier diversity professionals. While at Tuck (2015–2018), I continued writing a newsletter that was distributed to the MBE and supplier diversity graduates of the program.

After leaving Tuck, I could not shake the commentary bug and I started writing a column for the New Haven Register and Connecticut Post, part of the Hearst Newspaper group. This column appeared every other week from 2019 to the present. This column coveres MBE and corporate supplier diversity, but given that this was a more general audience, I covered other subject matter, largely based on what was taking place in American society and the world.

The third chapter includes commentaries that cover my economic analysis of markets with a particular focus on macroeconomic issues of economic growth, inflation, unemployment, and wealth and income inequality. Also included in this chapter I address the economics of markets ranging from banking to transportation.

The fourth chapter includes commentaries on the intersection of race and economics. Here I cover how race has played a central role in the development of America and continues to influence public policy.

Chapter 5 covers political action and political events that have dominated the news over the past 8 years.

Chapter 6 includes commentaries on health and education issues, particularly as they impact Black communities.

Chapter 7 includes commentaries on the criminal justice system and Black Americans' disproportionate engagement with the system.

The eighth chapter includes political and economic commentaries that do not fit easily into the previous chapters.

The final chapter is the chapter where I try to put the book into a political perspective by asking what all of this means. Specifically, I address strategies Black Americans (DSP) should consider as we enter the second Trump administration. The choices and strategies used for Black progress fall into three general categories: Full integration into American capitalist society; Separation from Whites; and Socialism. It would not surprise me if some readers might start at the end of the book or some other place. Fortunately, you can start anywhere in the book.

It was difficult to cull the commentaries to fit in the page goals of the book, and I strongly encourage readers interested in seeing all my published work to contact me, I would be happy to share what I can.

CHAPTER 2

Black Wealth and Black Minority Business Development

Abstract Entrepreneurship is one of the primary ways Americans build wealth for themselves and their families. Black American entrepreneurship has been hindered by the legacy of slavery that prevented slaves, who were property, from owning property. Combined with housing and employment discrimination after the end of slavery, Black entrepreneurs struggled. This chapter looks at current public and private policies that address this important issue.

Keywords Entrepreneurship · Wealth · Capital · Access · Diversity · Economics · Business · Black

2.1 What President—Can Do for Diverse Businesses

Fred McKinney, Ph.D

September 14, 2016

On January 20, 2017, a new President will move into the Oval Office. Over 15 million minority- and women-owned enterprises have much at stake in this election. Between now and Election Day, November 8, the diverse business community can clearly make the case for federal

F. McKinney, *Critical Issues Facing America*,
https://doi.org/10.1007/978-3-031-93217-5_2

policies that will have the most positive impact on the development of diverse businesses. While there is great diversity within the diverse business community, I want to suggest a few policies where there might be unanimity.

1. Expedited Payment for Diverse and Small Businesses

Under President Obama, the federal government improved upon the terms of the 1999 Prompt Payment Act. The PPA called on the federal government to pay small government contractors in 30 days of submission of an invoice on completed/delivered work. On September 14, 2011, the Office of Management and Budget issued Memorandum 11–32, "Accelerating Payments to Small Business for Goods and Services." The incoming president would be wise not to reverse this Executive action.

Fast payment to small contractors provides much needed working capital for firms often with limited access to working capital that larger firms take for granted.

In fact, the new president should expand prompt pay to all large federal prime contractors. All federal prime contractors should pay their small business suppliers within 15 days of product delivery or completion, regardless of whether the large contractor has received payment from the federal government. The pay when paid practices are not conducive to minority business development and should end at the federal, state, and local level.

2. Increased Access to Global Markets

The primary organization within the administration that can have a significant influence on minority and small business access to global markets is the Export–Import Bank. There was much controversy overextending the funding to this organization because critics believed it was a form of corporate welfare that benefited a hand full of large U.S. manufacturing exporters. Despite these objections, the EX-IM Bank is charged with growing U.S. small business access to global markets. This activity needs more resources and more attention in the incoming administration.

In addition to more resources, the EX-IM Bank needs to be considerably more strategic in its approach. For example, the EX-IM Bank could more aggressively partner domestic minority companies, who have natural connections to global markets, to larger non-minority companies. Why not connect more African American small businesses with African business opportunities; more Mexican American small businesses with Mexican business opportunities, etc.? If large non-minority U.S. companies were smart, they would be doing this as a matter of self-interest. The new administration can use the EX-IM Bank to make this more than just a pipe dream.

3. Expansion of Infrastructure Spending

The new administration needs to expand infrastructure spending with a target towards underserved and low-income communities. This infrastructure investment needs to include but go beyond the traditional roads and bridges.

The new administration should make as a goal that every inner-city (low-income) census tract in the country has unlimited wireless internet access; that every child in every low-income school district has access to a device that gives them access to the internet; and that ride sharing companies be required to serve low-income communities as well as they service other communities. Studies have shown that one of the major barriers facing low-income workers is their inability to access job opportunities outside where they live because they do not have cost effective transportation. These are the infrastructure investments that will make the minority and women business entrepreneurs of the future, and the employees of minority- and women-owned businesses today.

4. Create Incentives for Private Equity, Hedge Funds, Pension Funds, and Other Sources of Capital to Invest in Minority- and Women-Owned Businesses

There are trillions of dollars in capital markets that never come close to minority- and women-owned enterprises even though minority and women workers contribute to these funds through their 401Ks and tax dollars. We understand that the fund managers who control the investment decisions of this incredible amount of capital are chasing returns and attempting to manage risk. We also understand that for the most part, these managers are not looking to minority- and women-owned enterprises to generate those returns and manage risk. The new administration can use its "bully pulpit" to encourage pension funds, investment companies, hedge funds, and others to make more investments in minority-

and women-owned enterprises. This is likely to be done through intermediaries like the members of the National Association of Investment Companies and similar organizations.

5. Increase Training Dollars for the Development of Minority and Women Entrepreneurs

The new administration needs to more than quadruple the budget of the Minority Business Development Agency. This agency has been abandoned by Congress and needs to have the resources it needs to develop minority businesses. Currently, the MBDA is the smallest federal agency in the Department of Commerce and the only federal agency tasked with developing MBEs. It is time to consider expanding the MBDA to include the development of all diverse businesses. With 15 million minority- and women-owned businesses, a $30 million budget is not enough.

6. Use the Bully Pulpit to Encourage Corporations and State and Local Government to Utilize more Diverse Businesses

The new administration needs to make minority- and women-owned businesses and entrepreneurs much more visible. Part of this promotional effort should be directed at federal agencies. These agencies have people responsible for increasing utilization of minority- and women-owned businesses. They need the resources to do their jobs.

The new administration can also call upon the states and local governments to contract more effectively with minority- and women-owned businesses, particularly when there are federal funds being used on state and local projects. The administration is able to use its power to influence outcomes for millions of minority- and women-owned enterprises and it should do so aggressively,

7. Common Certification

The new administration needs to convene a high-level meeting of all public and private certification agencies and get them moving towards a common certification and certification clearing house. This could be done without threatening the historic organizations that have been so instrumental in the development of minority- and women-owned enterprises.

This common certification effort would save minority- and women-owned enterprises hundreds of millions of dollars. The money spent on inefficient certification regimes could be used in far better ways without diminishing the importance of minority and women business certification.

The new administration is likely to not be thinking about minority businesses as a priority. And can you blame them? In politics, if you do not ask—you will not get. It is time for us to ask, and demand that our interests as minority- and women-owned enterprises be addressed.

2.2 Mr. Trump and Black Unemployment

September 7, 2019

In the run-up to the 2016 presidential race, then-candidate Trump asked a question specifically directed at black voters: "What the hell do you have to lose?" Any black voter who answered such a deceptive question with a "nothing," in retrospect looks like a fool. Black unemployment rates have declined under Trump from 8.4% in January 2017 to 6% in July 2019. It would be deceptive of me and quite foolish if I were to claim that the continued decline in black unemployment rates is not good news. However, we must look at the historical context of the decline in unemployment rates and the other actions taken outside of changes in unemployment by this administration to accurately provide an answer to the question Trump asked in 2016 and will likely ask again in 2020.

The historical context is important because economic cycles do not neatly synchronize with U.S. presidential elections. As a result, presidents often receive too much credit for economic growth and too much blame for economic slowdowns. One exception to the inaccuracy of attribution is when the economy is clearly struggling. During these times, like those that existed when President Obama took office in January 2009, administrations can either prolong the recession with bad economic policy or shorten the recession with the right mix of fiscal policy. (The other major player in these situations is the very important role the Federal Reserve plays in countering economic downturns.) But presidents come into office at a particular time in the economic cycle that they had very little if anything to do with; they inherit the condition that they find. President Trump has often claimed that the U.S. economy he inherited in 2017 was in a mess created by his nemesis and predecessor, President Obama. This claim that the economy was in a mess and that "he alone" was going to fix it, is false in its premise and in its understanding of how the economy works. This is not to say that presidents and administrations don't matter. They do. However, let's review the facts.

When President Obama took office in January 2009, the U.S. unemployment rate was 7.8% and the economy was losing over 100,000 jobs every month and had been since September 2008. Beginning in October 2009, the U.S. unemployment rate declined steadily and consistently throughout the remainder Obama Administration because of steady, albeit slow economic growth by historical standards. Yet during the Obama years, black unemployment rates declined from 14.8% in 2009 to 8.4% in 2016. The fact that black unemployment rates continued to decline under President Trump is a testament not to him, but to the continuation of the economic growth started under President Obama.

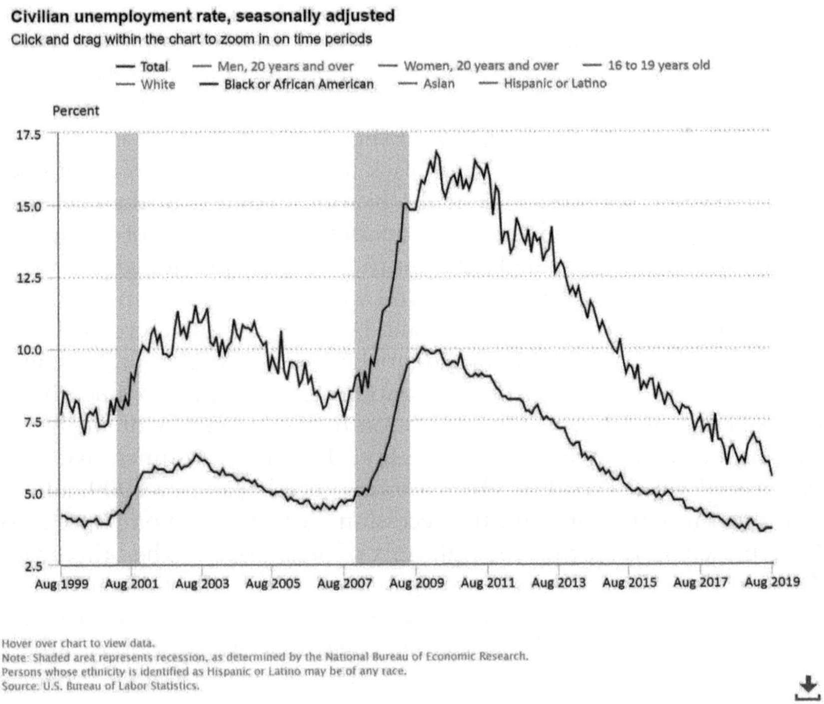

As we approach the 2020 election, President Trump, unlike candidate Trump, will undoubtedly claim that he has been net positive for black voters and black communities because of continued declines in the unemployment rate. You do not have to be Nobel Prize winning economist to read the chart above and conclude that that black unemployment rates have continue their decline started under President Obama.

The question of "What the hell do you have to lose?" is best answered not by focusing on the unemployment rate, but by looking at the impact having a White supremacist in the Oval Office. White supremacists believe that black Americans would exchange the rights and privileges of citizenship for lower unemployment rates. We should not have to choose and should not accept this indecent proposal. The doctrine of White supremacy is a cancer on the Republic. Jobs in the context of the social toxicity resulting from the words and actions of this president are not only beyond his direct control but are also a poor substitute towards the building of a more just country where we come together as Americans.

2.3 Black Business, the Wealth Gap, and the Case for Reparations

June 25, 2020

America is taking a collective course on American history this summer because of the George Floyd murder and the subsequent protests. One of the subjects covered this summer is the Tulsa Race Riot of 1921 where Whites destroyed the vibrant African American community known as Black Wall Street. Many Americans have also learned that memorials and statues of Confederate generals and leaders like Jefferson Davis and Robert E. Lee were not constructed before the Civil War or immediately after the Civil War; most were built in the twentieth century during the height of the Jim Crow era of government-sponsored American Apartheid. In fact, America was the model for South African Apartheid, and Nazi Germany. Hitler even acknowledged Jim Crow as a model for his vision of Germany. Americans have also learned that the last Black Americans were freed from slavery on June 19, 1865 (Juneteenth) more than two years after the Emancipation Proclamation. White Americans have learned that there is deep systematic bias by police departments across the country that have resulted in the deaths of thousands of Black Americans. However, one thing that has not been prominently discussed is the relationship between this history and its connection with the racial wealth gap and state of Black businesses.

According to the Federal Reserve, the typical White household has $171,000 in net worth while the typical Black household has $17,600 in wealth. Net worth is calculated as the difference between what you own and what you owe. This racial wealth gap manifests itself in lower rates of homeownership, greater levels of poverty that is passed down from one generation to the next. Yes, just as wealth can be passed from one generation to the next, so too can poverty. And because homeownership is one of the primary methods of funding a business, Black entrepreneurs typically start with significantly less capital, resulting in smaller unstable enterprises. A recent report showed that 41 percent of Black businesses have shut down since the start of the pandemic compared to only 17 percent of White businesses. The Minority Business Development Agency of the U.S. Department of Commerce estimates that only 4.2 percent (109,000) Black-owned businesses out of the over 2.6 million Black-owned businesses have employees. Over 21 percent of White owned businesses have employees. This is the direct result of the lack of household wealth in black communities.

Unfortunately, the racial wealth gap is not going to close “naturally,” in fact, it is only going to get worse. This is not a political statement, this is math. Assuming both Black and White household wealth increases by seven percentage points every year for the next thirty years, White household wealth will grow from $171,000 to $1.3 million by 2050, while Black household wealth will grow from $17,600 to $134,000. The racial wealth gap will increase from $153,400 to $1.17 million in 2050.

The only way to close this wealth gap and give Black entrepreneurs a chance to form businesses and for Black communities to have a chance to experience better schools, better healthcare, better housing, and better public services is for there to be a massive wealth transfer to Black households. In other words, it is time for a national conversation on the case for reparations.

Many opponents bristle at the word reparations. Some argue that they were not party to the crime of slavery and should not have to pay for those historical wrongs. Some state that is nearly impossible to identify who should be eligible for those payments. Some argue that reparations would only heighten tensions between Blacks and Whites. And some even argue it would not make a difference in the material conditions of Black Americans.

My brief response to each of these arguments is that if each of the 14 million Black American households were to receive $100,000 in reparations (for a total cost of $1.4 trillion), it could be completely financed by a one-time 3 percent wealth tax on the wealthiest one percent of Americans, who combined, own over $36 trillion of wealth. Therefore, 99 percent of Americans would not have to pay one cent to correct part of the damage caused by the history of slavery and Jim Crow. These resources would make a material difference in the lives of not only every Black American, but it would also drastically reduce if not eliminate the need for much of the current social welfare system that imperfectly addresses the problems of this wealth gap that do cost most White Americans.

In January of 1865, General William Tecumseh Sherman asked Black leaders in Savanah, Georgia what they wanted. Those leaders had a one-word answer—"land." General Sherman then issued Field Order Number 15 that authorized 400,000 acres of land from Savanah to Jacksonville, Florida be transferred to the newly freed slaves. Each former slave household was to receive 40 acres of tillable land. Unfortunately, Lincoln was assassinated on April 15 of that year and Andrew Johnson, his successor, rescinded General Sherman's order and gave the land back to the Confederate slaveholders.

If America wants to finally end the Civil War, we must provide reparations to Black Americans. Only then can we as a country move beyond race.

2.4 Why White Consumers Should Buy from Black-Owned Firms

7/24/2020

According to the American Heritage Dictionary, synonyms for black include soiled, evil, gloomy, sinister, and even grotesque. In contrast, synonyms for White include pure, unsullied, generous, and decent. Is it any wonder that black businesses have a branding problem?

These synonyms are cultural and Eurocentric artifacts. In many African languages the use of these words as adjectives is reversed with White indicating sickness, weakness, and evil while black means beauty and power. But English is our language not Hausa (a West African language spoken by 200 million people) and while the amount of melanin in our body does not determine intelligence, character, or humanness, it unfortunately has determined how we treat each other.

I bring this up because it has become increasingly clear to me that there is a national interest in promoting black business and to go beyond the typical entreaties to middle and upper-middle-class blacks to support black businesses. It is necessary now for everyone, particularly White consumers, to overcome the historical and etymological barriers that constrain black-owned businesses. If black businesses had the economic support of White consumers, we would not be in the mess we are in.

In my fifteen years as CEO of the Greater New England Minority Supplier Development Council (GNEMSDC), I worked hard to increase opportunities for certified minority owned companies to become suppliers to large corporations. We had our successes and our challenges, but over my tenure, we helped increase spending by large corporations with certified minority firms by hundreds of millions of dollars in Connecticut and the region.

Largely for this work, in 2017, I was awarded the Ronald H. Brown Leadership Award by the U.S. Department of Commerce's Minority Business Development Agency. Ronald H. Brown was the first African American to lead the Department of Commerce. Secretary Brown tragically died in an airplane crash in 1996 while conducting official business.

When I was invited by the incoming Trump White House to do a photo opportunity with the new president. I declined the invitation. The new administration had just released its budget which eliminated the Minority Business Development Agency. The irony of being presented this particular award by this president was too much given his intention to eliminate the only agency in the federal government designed to support black businesses. In retrospect, declining this invitation was one of my better decisions.

Recent data on who received Paycheck Protection Program loans reflects how much work we still need to do to support and build black businesses. The PPP did not require race or ethnic data on any of the borrowers for loans under $150,000. We know however, that of the over 16,000 loans made to Connecticut businesses, only 23 loans over $150,000 were made to black-owned enterprises. We also know from data collected by the MBDA in 2012 that only 109,000 of the over 2.7 million black-owned firms had paid employees. These numbers indicate that there is a problem facing large numbers of black-owned enterprises that no amount of "happy talk" will fix.

Why should White consumers care about black-owned businesses?

It is understandable why companies do not generally support their competition by telling their consumers to shop elsewhere. But this is not the issue. The reason why White consumers should support black-owned businesses is that their support creates jobs, opportunities, and wealth in black communities. With jobs and wealth come reduced demands on public dollars. With jobs and wealth comes increased economic stability. With jobs and wealth come reductions in crime, increases in home ownership, more fiscally stable cities and towns, better schools, and a more peaceful and just society.

The last reliable set of data measuring gross receipts for black-owned firms was done in 2002! Adjusting the gross receipts of black firms to 2020 prices would total about $129 billion in sales. American consumption in February 2020, pre-pandemic was $13.9 trillion. There are over 48 million black Americans, i.e., 14 percent of the U.S. population. If black firms had captured 14 percent of total American consumption spending, black firms would have had $1.9 trillion in gross receipts. The gap between what is and what could be is over $1.8 trillion. This gap will only be closed if White consumers make a concerted effort to support black businesses by spending a small fraction of their dollars with black-owned firms.

While growing black businesses is not going to immediately cure all social ills, we know from examples what it can do. One of the largest and most successful certified minority businesses during my time at the GNEMSDC was Specialized Packaging Group, owned, and founded by Carlton Highsmith. My position at Quinnipiac University was funded by Mr. Highsmith and People's United Bank. Mr. Highsmith was the primary investor in the Connecticut Center for Arts and Technology (CONNCAT) in New Haven. CONNCAT is now one of the most effective nonprofit organizations in the country. Minority business success benefits everyone.

During this time of grappling with the challenges of racial justice, I have heard from many of my White friends who have asked me: "What can they do?" What they can do is support black-owned businesses. Black-owned businesses are everywhere and in every industry. You just have to look for them.

Now, I think I will give Thelma's Restaurant a call and order my Sunday dinner of sweet potatoes, collard greens and the best mac and cheese in the state.

2.5 The Case for Raising the Federal Minimum Wage

February 18, 2021

Earlier this week at a town hall meeting in Milwaukee, President Biden expressed his support for an increase in the federal minimum wage from its current level of $7.25 per hour to $15 per hour. Biden said, "No one should work 40 hours a week and live in poverty." This increase would be phased in over several years.

There are two schools of thought on this issue. Both are armed with data and studies that seem to indicate support for their conflicting positions. But there are some things that those studies agree on even when they may not agree on the effect changes in the minimum wage would have on the number of American workers helped, who would be helped, and the number of Americans harmed.

First it is important to establish some facts. (It is refreshing to have policy debates based on facts and not just pure ideology.) In the fourth quarter of 2020, the median earnings for all U.S. men working full-time was $1069 per week and $896 per week for women. This amounts to approximately $28 per hour for men and $23.89 per hour for women. The weekly earnings for White men averaged $1058 per week compared to $778, $1457, and $769 for Black, Asian, and Hispanic men, respectively. White women earned $859 per week compared to $738, $1008, and $654 for Black, Asian, and Hispanic women, respectively. By comparison, the weekly earnings of a full-time worker at the current federal minimum wage are $290 per week. An increase to $15 raises it to $600 per week, still significantly below current average earnings.

We also know that most minimum wage workers today are young. While only one-fifth of workers are under 25, this group of young workers makes up over 50 percent of the approximately 2 million workers making the minimum wage. This informs us that a potential compromise might be a tiered increase in the minimum wage whereby young workers will have a lower minimum than adult workers. But even given the higher concentration of young workers making the minimum, 50 percent of minimum wage workers today are prime-age adults, presumably with adult responsibilities. I do not know how any adult can comfortably and independently survive making less than $1200 per month anywhere in the United States.

The biggest impact on raising the minimum wage would be on increasing wages for those who are already making more than the minimum wage. Economists think of wages as a structure, with the minimum wage at the foundation of that structure. When the minimum wage increases, all other wages are also likely to increase to maintain the relative differences. Workers making $8/hour would also see their wages increase as the new minimum exceeds their wage. Therefore, an increase in the minimum would impact millions more workers than the 2 million currently at the minimum.

We know that rural and Southern states have more workers working at the minimum wage compared to states with significant urban/suburban populations. The minimum wage in Alabama, Mississippi, Idaho, Iowa, Kansas, Kentucky, Louisiana, New Hampshire, North Carolina, North Dakota, Oklahoma, Utah, Wisconsin, Virginia, and Wyoming are all at the current federal minimum wage of $7.25. A move to raise the federal minimum wage would have no direct impact on Connecticut employers because Connecticut is already at $12 per hour and scheduled to reach $15 per hour in 2023, probably long before that rate is reached nationally.

Those who argue against the increase state, with some support from academic studies, that an increase would create more unemployment among low-wage workers. This argument is based on classic economic theory that states that the demand for workers is a function of how much economic value they create for their employer. For example, if a worker can generate $12 of economic value in an hour's worth of their time for the employer, and the worker's wage is less than $12, the employer has a profit incentive to hire that worker. Alternatively, if that worker's wage goes to $15 per hour and their economic contribution does not change, the employer will let that worker go. Therefore, conservatives argue that raising the minimum wage can harm low-wage workers whose productivity is lower than the new minimum. Secondarily, conservatives argue that small business owners who are more likely to hire and need low-wage workers would be harmed because of higher costs associated with the higher minimum wage.

For Connecticut businesses and workers, this debate about an increase in the federal minimum wage is a moot point. There is, however, one important impact an increase in the federal minimum wage could have on Connecticut businesses. If these other states are forced to raise their wages, almost immediately, Connecticut becomes relatively more attractive to businesses nationwide. Connecticut minimum wage workers are more productive, even considering cost of living calculations, than minimum wage workers in the states at the current minimum. Connecticut politicians and, importantly, business leaders should welcome the increase in the federal minimum because of the influx of employment that could come into the state.

Even if I lived in one of the low-wage states, I would argue for an increase in the federal minimum wage. I would also argue that, like Social Security payments, there should be built-in cost of living increases (or decreases) to the minimum based on changes in inflation. If future increases were automatic, we would eliminate the political problem raising the rate creates.

Even in these low-wage states, we have the income and wealth to share with hard-working Americans. We have seen during this pandemic that we can lift millions of Americans out of poverty by providing them with enhanced unemployment insurance payments and child care credits. Why not lift millions out of poverty by paying American workers a living wage? They are more than worth it.

2.6 The Cost of Corruption to Minority Business

October 17, 2022

Last month, a North Haven, Connecticut HVAC firm was found guilty in a bid rigging case. The firm Langan Insulation was found guilty of conspiring with other installation firms to rig bids on universities, hospitals, and other companies in Connecticut. The guilty company will have to pay close to a million dollars in fines and penalties, and the co-owner will spend a year and one day in a federal penitentiary thinking about his crime. Five other participants in this case pleaded guilty.

The New York Attorney General, Letitia James announced that her office had secured a $125,000 judgment from the Pike Company, a Rochester, NY, construction firm that falsely certified that it complied with state diversity requirements when it subcontracted the supply of materials to NY State certified M/WBEs when in fact it had done no such thing. Pike was one of the TEN companies that were found guilty engaged in fraud associated with NY's supplier diversity program. This time the client was Rochester Public Schools.

Cases like this undercut and neuter even the most well-designed M/WBE programs. The Connecticut case involved contracts with private sector and public sector entities. The NY case was defrauding the citizens of Rochester. In both cases, non-minority firms got together to collude to secure contracts or increase the amount of money they pocket on a particular contract.

Think about the damage done to legitimate M/WBEs when this kind of behavior is taking place. These are just two cases. I believe, admittedly without hard evidence, that this is just the tip of the iceberg—the part that we can see. It is no telling how many of these shenanigans are taking place in communities across the country in public and private sector contracting. But rest assured, these are not the only two cases of fraud and corruption involving minority business programs.

It is difficult to estimate the financial harm to M/WBEs from fraud and corruption, but the cost goes far beyond financial. M/WBEs who do not win contracts because of fraud, experience slower growth, lower sales, lower profits, and slower employment growth.

On a recent consulting engagement with an Ivy League university, the client was pleasantly surprised that diverse businesses won a disproportionate share of contracts when opportunities were competitively bid. Minority businesses can compete when they are given a chance—when the playing field is relatively flat. There is no reason to believe that MBEs cannot compete with the big boys when given the chance. But corruption is a way to keep the field slanted in the favor of non-minority corrupt businesses.

It gives me comfort that the U.S. Department of Justice and some State Attorneys General are paying attention to corruption on public and private contracting. There are laws that allow whistleblowers to share in the financial settlements reached between offenders and the courts. Why not share the proceeds of these crimes with the minority businesses who are directly harmed by these practices? and expose the criminals to appropriate authorities?

The fact that non-minority firms are conspiring with each other, and in some cases, the buyers with large public and private organizations should be shocking. Unfortunately, it is not. We as minority business advocates cannot ignore this criminality. We must try to eliminate corruption and abuse of M/WBE programs. The money that is illegally "won" by these criminal firms is money that could have gone to legitimate M/WBEs.

2.7 Ten Excuses for Not Doing Supplier Diversity Well

December 12, 2022

No corporate CEO or corporate communications leader in a post George Floyd world would dare to publicly make the excuses I am about to delineate, but they are things I am still hearing and reading between the lines when I speak to corporate leaders on the topic of suppler diversity. The fact that I am hearing them probably means some of you might also be hearing similar things from colleagues and others within corporate North America. For some time, I have been cataloging these objections and sources of negativity. As any good salesperson knows, it is better to anticipate and be prepared for the reasons why a potential customer will not buy something than wait until you hear it for the first time; by then it might be too late. An unplanned response to an objection might not be as powerful as one that is anticipated. Selling supplier diversity is like selling just about anything that a corporation does not already own—it takes work and planning.

My list is not exhaustive, but here are the top ten excuses I hear about why corporations and organizations do not do a better job at supplier diversity. It is important to remember that these are the things well-meaning companies say, think, and more importantly do.

1. We Do Not Have the Budget to Implement an Effective Supplier Diversity Program

Corporate supplier diversity is not free. At a minimum, you will need to make someone or a group of employees responsible for a program to increase opportunities for diverse and previously underrepresented groups of businesses. You will need to make some changes to the data you collect on current and future suppliers. You will need to review your contracts with future suppliers to extract certain language and include new language designed to increase fairness. You will need to train buyers in new processes to increase inclusivity. There are costs and changes that supplier diversity requires. If your company does not do these things now, doing them will require resources. Supplier diversity should be no different from anything else a company spends money on—costs must be viewed with benefits associated with those costs. Incurring supplier diversity costs has benefits. The benefits must be as well articulated as the cost.

2. We Do Not Buy Goods and Services that Are Supplied by Diverse Businesses

Regardless of the industry, there are some goods and services within a corporate supply chain that are supplied by a limited number of suppliers. Economists call these markets oligopolistic markets. For example, 97% of Deep Sea Passenger Transportation services are controlled by the four largest companies and 96 percent of Home Centers are controlled by the four largest companies. You probably know the names of the companies in these highly concentrated industries.

Some highly concentrated markets of suppliers are that way because of significant financial, technical, or legal barriers to entry. Some oligopolistic markets develop because of extra-legal barriers to entry like collusion, discrimination or threats of violence. Regardless of how a market became oligopolistic, chances are there are still wide swaths of goods and services that are supplied by diverse businesses. Taking the time and effort to seek them out and welcome them to compete for your business is key.

3. We Tried Working with Diverse Suppliers and They Are High-Cost and/or Lower Quality

The reality or perception that diverse businesses are higher cost and lower quality is sometimes associated with a limited sample of experiences with diverse suppliers. It is important to distinguish between high-cost suppliers and low-quality suppliers. Any company that is both higher cost and lower quality is not likely to be in business for long regardless of the race and gender of their owner. Higher cost diverse firms exist as do higher cost non-diverse firms. Low-quality diverse firms exist as do low-quality non-diverse firms. It is counterfactual to assume all diverse firms are both excessive cost and lower quality firms. And even if your company, or more likely the buyers from your company, had a bad experience with a diverse supplier that should not reflect on all diverse suppliers. Similar bad buying experiences with non-diverse companies do not usually impact how buyers view other non-diverse companies, but this is not the case with how diverse companies are treated. Economists call these types of behaviors externalities where the actions of one business impact the success of other businesses. I call it lazy thinking.

4. We Need Suppliers Who Understand Our Company and Our Industry

Incumbency has its advantages, ask any U.S. Senator running for reelection. Just as in politics, familiarity helps. The problem diverse firms have is that there is a history of not being included in corporate supply chains. This is a self-fulfilling loop of doing the same thing and attributing the lack of success of new firms, particularly new diverse firms to their inability to break into this loop of exclusivity. Chances are, firms who use this excuse are not looking hard enough for diverse firms who do have experience in their industry or similar industries.

There are risks associated with changing suppliers. Like the issue of costs, risks must be evaluated along with the benefits associated with making changes. What many companies do not factor into this thinking is the risks of staying pat with the same suppliers who "understand" their company and their industry. You might be paying too much for that understanding.

5. We Don't Sell to the Public Sector, so We Do Not Need a Supplier Diversity Program.

Modern supplier diversity started with the federal government because of the 1960 urban riots. The federal government put in place several executive orders designed to increase federal purchasing of goods and services supplied by minority firms. From that beginning, corporations many of whom were large federal contractors or large consumer product companies began implementing supplier diversity programs to address social ills and to stay on the good side of the federal government who also started asking large private companies to comply with federal rules and regulations. This notion that because a company does not sell to the federal government as an excuse for not having a supplier diversity is anachronistic. It might sound like a good excuse, but what it demonstrates is ignorance.

6. We Do Not Hear from Our Stakeholders that Utilizing Diverse Suppliers Is Important to Them

This is an excuse that likely reflects a limited perspective and or a lack of diversity among those who are considered stakeholders. Group thinking is common in leadership circles. There is plenty of scientifically supported literature on the dangers of not having diversity of thought when important decisions are being considered. These studies show the lack of diversity often leads to limited perspectives and less creativity when solving problems. Echo chambers exist in politics and business, and both are dangerous.

7. We Are Trying to Reduce the Number of Suppliers in Our Supply Chain—Not Increase the Number of Suppliers

Strategic sourcing has for decades informed corporate procurement of the advantages of leveraging suppliers by offering larger shares of the company's budget in exchange for lower prices. This has led to supply chain consolidation—i.e., fewer suppliers. Larger corporations also calculated the cost of having many suppliers within their internal systems. Reducing the number of suppliers as a sourcing strategy seems inconsistent with supplier diversity that tells a company to consider new suppliers. It is possible however to have more diversity in the supply chain and have fewer suppliers.

8. Switching Suppliers Is Expensive

While switching suppliers is and can be expensive, we also know that a lack of competition can be even more expensive. Some companies have dealt with this by requiring existing suppliers to avoid having their products and services put out for bid by having the suppliers guarantee annual reductions in prices. In these cases, the suppliers have to offer price discounts every year just to keep the business. Suppliers hate this but are often willing to sign up for these negotiated price reductions to keep the business. But most companies do not build in negotiated price reductions for multi-year contracts and this is harder to do now given the inflation we are experiencing. In the absence of negotiated price reductions, staying with suppliers can also be expensive. We have seen in our practice clients typically save money and increase diversity by going to the market for new suppliers.

9. Some Diverse Suppliers Are Not Diverse

It is true that there are diverse front companies and there are legitimate diverse companies who do not have any more diversity on their staffs than do their non-diverse competitors. These are two different types of problems, but both can impact decisions and perspectives. The issue of minority fronts is best dealt with by having effective certification. Third-party certifying organizations police diverse fronts. That is why businesses should only count spending with certified firms as diverse and should encourage known diverse firms to get certified to eliminate the demand for diverse front companies. The issue of diverse firms not having diverse staff is an issue that is often beyond the control of the diverse firm. Diverse firms rely on the same labor markets non-diverse firms rely on, and for too many of these markets the supply of diverse talent is not as deep as it should be.

10. We Prefer to Use Local Small Businesses

Preferring local is a good thing. It is however not inconsistent with increasing the utilization of diverse firms. Companies make tough choices every day. Choosing a local diverse small firm over a local non-diverse small firm can be an easier political decision than choosing a non-local diverse firm over a local non-diverse firm. I have also increasingly seen instances where large diverse firms are competing against small non-diverse local firms. It is a good thing that diverse firms are both local and non-local as well as small and large.

Supplier diversity is not easy, but it is not impossible, and done well, it pays dividends. I would love to hear from you on objections and resistance you have heard and seen in your work in corporate procurement and corporate supplier diversity. I am certain of very little in life, but I am certain, I have not captured it all.

October 28, 2023.

2.8 Historical Effects of Racism on Black Business

This past week I was a co-presenter in Baltimore to a group of 20 procurement and supplier diversity professionals from large corporations in the hospitality industry. Globally, this is a multi-trillion-dollar industry that includes airlines, hotels, and the food industry. The attendees at this six-hour training session learned how to improve their corporate supplier diversity program.

I started my talk with the story of Bruce's Beach. In the first decade of the twentieth century, Charles and Willa Bruce set out from Washington, DC towards California. They stopped in Albuquerque, NM for two years before making it to their final destination in Manhattan Beach, California, around 1912. Charles Bruce secured a job as a pullman porter. At that time, pullman porter positions were middle-class jobs for Black American men; they received steady paychecks for their work serving train customers.

Willa Bruce was an entrepreneur, who instead of waiting for her husband to return from his duties serving the travelers from Los Angeles to Salt Lake City, purchased a small plot of land directly on the Pacific Ocean in Manhattan Beach. She was the first Black American property owner in this beach community. She opened a small resort where Black beachgoers could change (and in some cases, rent swimwear), eat, and have an enjoyable time. In 1912, Jim Crow practices prohibited Blacks from going to some Southern California beaches. As a result, Willa Bruce's resort became a popular destination for Blacks throughout the region.

Things were looking up for Willa and Charles Bruce. By 1920, because of her success with her first property, Willa Bruce purchased a second property nearby. Following that, other Blacks started buying property near Bruce's Beach.

Willa Bruce's success, however, was viewed by some White Manhattan Beach residents, business owners, and politicians as a terrible thing. Seeing Black excellence and entrepreneurial success attracted the racist ire of the White town council which, with the stroke of pen, claimed eminent domain and forced the Bruces and the other Black property owners from their land to build what they claimed would be a public park. Interestingly, the overt racism by the town was memorialized in written communication that specifically stated the reason for the eminent domain action was Bruce's race. Contemporaneous memos used derogatory names for the Bruces and their African American customers. After the Bruces' ouster, it took over three decades for the town of Manhattan Beach to build that public park. The use of eminent domain was obviously just a ruse to force this successful Black business to close its doors.

I told this story to the hospitality industry because while Bruce's Beach opened its doors, Conrad Hilton opened his first rooming house in New Mexico. The historical contradictions between the two hospitality companies opening at the same time are striking: the Black firm was forced out of business despite having a prime location and a strong customer base, while the White firm went on to become the largest hotel chain in America.

I learned more about this case from the attorney who represented the descendants of Charles and Willa Bruce. The case of the Bruces was known, but it took the murder of George Floyd in May of 2020 to ignite a legal movement to right this historical wrong. Los Angeles County, which owned the land taken from the Bruces, settled with the Bruces' descendants for $20 million in 2022.

While it took 100 years for justice to happen, something remains unjust about this case. It is not an unrelated fact that in 1921, White mobs burned the Greenwood section of Tulsa Oklahoma, an extremely successful enclave with hundreds of successful Black businesses. It was so successful that it became known as Black Wall Street. Racial pogroms against Black Americans took place across the country from 1917 to 1923, having devastating consequences for Black business communities in cities including Houston, New York City, East St. Louis, Chicago, Washington, DC, Rosewood, Florida, and Charleston, South Carolina.

This history is important to note because Black businesses, as distinct from "minority" businesses, had setbacks that were uniquely American and historically devastating. In fact, the special case of Black businesses is a reason there should be special consideration for Black businesses.

The Hilton family is one of the wealthiest families in the world today because of the success of the Hilton Company. I harbor no ill will towards the Hilton Company, Conrad Hilton's heirs, or their success. But think about the Black generational wealth that was thwarted or never created because of racism like that experienced by the Bruces and countless other unnamed Black families.

Not all Black businesses today are struggling, but they are typically significantly smaller and have less access to capital than other minority-

and women-owned businesses, not to mention even larger differences compared to White owned small businesses. But there is something that can be done to correct this disparity. First, corporations and governments at all levels need to acknowledge this unique history and its deleterious and continuing consequences on Black businesses and Black communities today. Secondly, corporations and governments need to promote and support Black businesses by buying from them boldly and unabashedly. Even if they do not want to pay reparations, they can—without detriment—do business with Black businesses.

Unfortunately, Willa and Charles Bruce died in poverty. But thanks to some good people in LA County government and some great legal support, the Bruce's heirs will be able to leave a financial legacy to their children and grandchildren.

2.9 Minority Supplier Diversity Is Dead

November 14, 2024

Thank you for that warm introduction. Let me thank Shaunne Thomas of Guilford County who gave me an opportunity to conduct an economic impact study of the MWBE Program for Guilford County, North Carolina. Let me also thank the MWBE Coordinators Committee for inviting me to speak today. And I want to thank all of you for taking the time to be with us today.

The timing of this meeting is most propitious. I was asked to attend this session and speak on the subject of the economic impact of supplier diversity about two months ago, before the presidential election that happened just a week ago yesterday. I started preparing my remarks before the election, all aglow with the fervent hope and expectation that Kamala Harris would prevail in her historic attempt to succeed President Biden. Had my hopes been realized, my comments would have been suffused with a cautious optimism and sanguinity about the prospects for your continued success in the field of MWBE development and support—because America had a president in Kamala Harris who was willing and able to put the full power of the federal government behind what we are all so passionate about.

But against the common sense of an electorate driven by fear and ignorance, my hopes, and I suspect your hopes were dashed like so many fallen dead leaves on a cold November day.

This talk is propitious because the political environment and the context in which we do our work has changed dramatically and perhaps permanently. I am forced to admit the painful truth that minority supplier diversity, as we know it, is dead. You could even conclude that election of Donald Trump means the final nail is in the coffin of diversity in general.

I come to this conclusion not because I desire it to be true. I arrive at this belief because in a struggle that started before, we were born with goals of justice and equality, the tactics and strategies must confront the realities of the situation.

The life of supplier diversity lasted a little more than 50 years. March 5, 1969, was the birthdate of Minority Supplier Diversity with President Nixon's Executive Order 11,458 establishing the Office of Minority Business Enterprise. This Executive Order also supported the development of the National Minority Supplier Development Council. While Nixon issued the Executive Order, it was Black Americans who fathered the movement through peaceful and not so peaceful civil disobedience in the 1950s and 1960s.

Minority Supplier Diversity's life was a hard life full of periods of growth interspersed with setbacks. But now what makes this different is that Supplier Diversity does not have the support of the incoming Trump Administration. If the past is prologue, the new administration is likely to immediately cancel all of the previous Executive Orders designed to assist minority businesses. I have not forgotten Trump's first budget proposal, he called for the elimination of the Minority Business Development Agency, the only agency within the federal government designed to assist minority firms. He was thwarted in this attempt by a Democratically controlled Congress. Now the federal government is under the control of a Republican Congress that fully buys into the Trump agenda. Trump's recent announcement that Elon Musk and Vivek Ramaswamy will lead the effort to pare down the federal government will likely pick up where he failed in his first term. Trump's fragile ego will not allow any program to survive that benefits groups he believes did not support him; and the most visible group of non-supporters are Black Americans.

The Supreme Court and the lower courts are not supporting Supplier Diversity as demonstrated in several of their recent rulings on Affirmative Action, the U.S. Department of Transportation's Disadvantaged Business Enterprise program and other efforts that they view as violating the rights of White Americans. And with the new administration comes the power to nominate new judges and Supreme Court justices who agree with these views. There is a very real possibility that the conservative majority on the U.S. Supreme Court will grow from 6–3 to 7–2 with the impending retirements of Clarence Thomas, Samuel Alito, and Sonia Sotomayor. It has recently been suggested that Justice Sotomayor should retire now so that President Biden can appoint a Justice who is ideologically in synch with Sotomayor. But it is an open question whether the Senate would allow such a late change even though they used the same tactic to appoint Justice Amy Coney Barrett on October 27 just a week before the last presidential election. The hypocrisy of that decision considering Senator Mitch McConnell's delay of President Obama's nomination of Merrick Garland to the Court, 8 months before the 2016 election on the grounds that the next president should make that choice is astounding. The mercurial McConnell will come up with some expedient rational for not supporting a late term appointment by President Biden. With a 7–2 majority on the Supreme Court, programs designed to ameliorate historical racial wrongs are not likely to survive.

The corporate community has already demonstrated its willingness to go with the anti-diversity flow with several companies announcing reductions in their support for supplier diversity and affirmative action. It seems that the post George Floyd corporate promises have been forgotten. In this new environment, there is little reward for "doing the right thing." I recently co-authored a White Paper for the Conference Board, an organization of the largest American companies. The purpose of the paper was to give its members a sense of what corporations were doing in this post—Students for Fair Admissions (SFFA) v. Harvard environment. In that landmark case, racial preferences in college admissions were ruled unconstitutional violations of the rights of White and Asian students. Our Conference Board study revealed that corporations were still saying they were committed to minority supplier diversity, but they did not like the political baggage associated with the word diversity. Some companies, to continue their efforts, were rebranding their programs as having a positive economic impact on local communities where they are located.

Many states and municipalities have also cut back their support for Supplier Diversity feeling those same cold political winds.

Even the nonprofit sector as demonstrated by recent lawsuits against the Fearless Fund whose mission it is to assist Black women entrepreneurs with their capital needs have been "successfully" attacked by opponents of diversity on the grounds that they violate the rights of White women. This was not just a shot over the bow; it was a chilling threat against Community Foundations, Corporate Foundations, and other non-profits designed to assist Black and other minority communities. Now these attackers will have the explicit support of the U.S. Justice Department, the Courts, and the Legislatures.

Minority Supplier Diversity now only has the support of the would-be beneficiaries—MWBEs and the warriors in this room. A program without supporters is not a program. This is why I conclude that minority supplier diversity is dead.

But all is not lost. After this long preamble, there is a solution. At least in the short-term, we must refocus the rationale for minority supplier diversity away from racial justice to community economic impact.

Economic Impact measures the effect of money spent on the broader economy. This methodology began with the work of John Maynard Keynes who developed the concept of the economic multiplier. Keynes' focus was on the overall macroeconomic impact of government and investment spending on national employment. The economic system is like a complicated machine where government, businesses, workers, households, investors, and foreign interests are connected to each other. Keynes demonstrated that in times of economic distress, government money spent on infrastructure can have a multiplier effect on the overall economy.

Community economic impact looks at how county government spending with MWBE firms impacts employment, income, philanthropy, and local government tax revenue and the overall health of the local economy. County economies are a microcosm of the country's economy. The same principles apply to your county as they do to the overall economy. Your spending can have a multiplier effect on the communities you serve.

The methods I use to calculate economic impact starts with a survey of MWBE firms. We need to know what these firms buy, who they buy it from, what they do with their profits (assuming they have profits), who they employ, and where they are located. We then need to look at what the counties buy, who they buy it from, where their vendors are located, and information about how the counties raise their operating revenues through various taxes.

With this information, we can begin to measure the community economic impact of money spent with MWBE firms. It sounds easier than it is.

Getting a representative number of firms to respond takes persistence and help from MWBE coordinators like you to impress upon the MWBE community that community economic impact analysis is what will keep money flowing to their firms, assuming the results are positive.

The positive results can show what a dollar spent with MWBE firms does for employment, tax revenues, philanthropy, income, consumption, and general economic health in counties across the state. One of the findings I have consistently found is that the MWBE community are some of the largest philanthropists. It makes sense given that many MWBEs are the wealthiest members of the community. Another important potential finding is that MWBE program expenses are good county investments. We might be able to show that the tax revenues generated resulting from the income and purchases made by MWBE firms more than pays for the cost of running MWBE programs. In an era of heightened fiscal scrutiny, this is a finding I am sure anyone in this room interested in their own job security and the survival of their programs would find critically important.

Without getting into the nitty–gritty of EI analysis, it is common to distinguish among the three main sources of economic impact. Firstly, there is the direct effect of your programs. A dollar spent with MWBEs is a dollar that those MWBEs now have that they would not have otherwise had. This is a direct impact.

Economic Impact does not stop there. The dollar earned by the MWBE also is spent. Economists have long recognized that one person's or firm's income is another person's or firm's expense and vice versa. The MWBE is now in a position to use that dollar earned from the County to pay the wages of its employees, purchase the services of other firms, pay taxes, and support the firm's/entrepreneur's philanthropic interests. These are the indirect effects which start their own multiplier processes.

The induced effects of this process involve the spending of employees who earn income from the money generated by the initial purchases by the County. Employees are wage earners and consumers. As consumers, their expenditures also support the local economy.

The sum of the direct, the indirect, and the induced effects represent the total economic impact we measure in this analysis. From that total economic impact, we can then calculate rates of return to the MWBE program. The ROI on MWBE spending is a very important figure to know, particularly when budgets are determined.

Economic Impact analysis takes the focus away from race-based programs and is consistent with the original purpose of the Nixon interventions. Nixon wanted MWBE development not because he wanted to increase the number of wealthy Black entrepreneurs. He pursued these programs as a conservative alternative to welfare which he and other conservatives viewed as ineffective and inconsistent with their capitalist ideals. "'If only the Blacks could support themselves, then we would not have to" was their reason for supporting these programs. It seems we have now come full circle.

Community economic impact also has the benefit of providing incentives for MWBE firms to hire workers in distressed communities, to buy from other MWBE firms in the community, and to support other local government and philanthropic goals. Community economic impact is also consistent with the development of communities that are economically self-sufficient and thriving.

On a personal note, I am in the process of moving to Prince George's County in Maryland from where I have lived in Connecticut for the past 48 years. I am familiar with my new home which is a majority Black County and also one of the wealthiest Black majority counties in the country. The newly elected United States Senator for the State of Maryland, Angela Alsobrooks was formerly the chief executive of Prince George's County. Additionally, Maryland has a Black governor, the mayor of its largest city, Baltimore is Black. And the mayor of nearby Washington, DC, is Black. This is the state where Black entrepreneurs, business professionals, and movers and shakers from across the county should move. The connection here between economic power, racial concentration, and political power is an undeniable lesson that community economic impact can be a powerful lever for change.

I want to close with some things for you to consider in these changing times. What should you be doing now? My first suggestion is that if you have not done it already, you should be taking a census of MWBE firms in your community. This census should not be of the MWBEs you are doing business with or the ones that are certified by your organization. Those are good places to start, but the census should cast a wide net. Community Economic Impact will require accurate and timely information on who and where the MWBEs are who might be affected the most by a Community Economic Impact strategy.

Secondly, you need to conduct a Community Economic Impact study that demonstrates the value of money spent with MWBEs for the taxpayers of your county.

We also need to educate the community on business and the economy. This effort should start with the children in our community. Economics and business education should be at the core of our education system. The curriculum in our schools can no longer afford to not teach our children at the earliest possible age how our economy works and how they can establish themselves as business owners and professionals. We owe it to them.

And finally, I want to stress that while MWBE programs in their traditional sense, are dead, it does not mean we should abandon our allies or disband our ability and willingness to work together for common goals. The end of minority supplier diversity does not mean we surrender. It means we retool and recommit to achieve an America where we all have a share and a critical role in its further development and future.

If I had additional time, I would share with you my thoughts on other strategies that are natural extensions of what I have shared with you today. Those strategies differ significantly from the MWBE strategy we have used to date to grow business opportunities and create entrepreneurial wealth. But I will leave that important discussion for a later time.

Thank you. I would be happy to answer any questions you have and we have time for. It has been my pleasure to be with you, and I hope we can continue to support each other in our mission to increase economic growth in all our communities.

2.10 How to Save the NMSDC

December 1, 2024

It was about ten years ago that I drafted a new business model for the NMSDC. The problems the NMSDC are facing today are not very different from the problems the organization faced 10 years ago or 20 years ago. The problem is the organizational model of the NMSDC is not consistent with the best roles for the component parts of the organization.

It is important to remember that before there was an NMSDC, there were local councils, beginning with the Chicago Minority Purchasing Council that started within the Chicago Urban League in 1968. It was after the Chicago Council model spread to New York and other major metropolitan areas that the NMSDC was formed in 1972. The NMSDC was originally funded with federal dollars made available in a Nixon era Executive Order 11,458.

Section 2 of the order stated:

> Promote the mobilization of activities and resources of State and local governments, businesses, and trade associations, universities, foundations professional organizations and volunteer and other groups towards the growth of minority business enterprises and facilitate the coordination of the efforts of these groups with those of federal departments and agencies.

The first major transformation of the NMSDC was to transform from primarily a federal government funded organization to primarily a corporate funded organization under the leadership of Harriet Michel, who not incidentally came out of the Civil Rights struggle and the Urban League organization.

Under Ms. Michel's leadership, the network of regional councils grew to over 40. Currently, there are 23 regional councils. Some of this reduction is the result of consolidations and mergers of councils. As one example, I led the consolidation of the New England Minority Supplier Development Council and the Connecticut Minority Supplier Development Council in 2008–2009 to become the Greater New England Minority Supplier Development Council.

There was always a struggle in the NMSDC network with the distribution of money because of its structure of national corporate members and local corporate members and to a lesser extent large MBEs labeled "Corporate Plus" MBEs and the other 99 percent of certified MBEs who made up the network. During most of the Michel and post-Michel era, the regional councils were supported by local corporate members, certification fees, events like trade shows, annual awards dinners and golf outings and some foundation support along with some distribution of national dues paid by national corporate members. National corporate dues were allocated based on whether the national corporate member had business activities in the regional council's area and therefore, might need the support from that local council to identify certified MBEs from that local area. Importantly, national corporate dues were determined by the size of the national corporate member's national business footprint. Large corporations like Walmart (who recently announced it is getting out of the diversity business) paid more in national dues than a large regional utility company. The larger national members had more funds to allocate to the regional council based on support received from the regional councils.

The NMSDC received its support from national corporate dues that were not distributed to the regional councils, and events like the national trade show and the national awards dinner. It is important to also note that up until the end of the Michel administration, the NMSDC and the hosting regional council shared in the risks and revenues from the national trade show. Now that event is exclusively an NMSDC event.

This system is fundamentally flawed because it is the regional councils who engage with both national and local corporate members the most and with certified MBEs the most.

The alternative structure I recommended over ten years ago called for a national office that was responsible for network strategy and policy, national communications, interactions with the Federal government, legal strategy, and international strategy. The national office in this model would be supported by the regional councils who would be the recipients of all corporate dues and all MBE fees. There would no longer be a distinction between local and national corporate members. A NMSDC corporate member would be an NMSDC corporate member. And there would be no distinction between a local MBE and a national MBE. There would just be NMSDC certified MBEs. The NMSDC board would be comprised of supporting corporations, MBEs, regional council leaders, and community advocates. Surprisingly, today there is no role for community advocates on the NMSDC board or the boards of the regional councils. This lack of community connection is a serious oversight and puts the entire network in a vulnerable position.

The regional councils would agree to a uniform fee structure for all corporate dues and MBE certification fees, eliminating incentives to "shop around" for membership and certification.

A further and important change I recommend is that the regional councils concentrate on large metropolitan areas and stop the charade of state and multi-state services. During my time as President and CEO of the Greater New England Minority Supplier Development Council, I visited Maine once in those 15 years. There are 40 Standard Metropolitan areas that have between 20 million residents (NYC) and 1.6 million residents (Milwaukee). The NMSDC should focus entirely on these SMSAs that have a combined population of over one half of the entire U.S. population of 330,000,000 people. Those SMSAs should define the regions.

A focus on SMSAs also makes possible to focus on local and community economic development. The organization has a legitimate goal of serving corporations supply chain needs, but the bigger goal should be community development through minority entrepreneurial development. Corporate minority supply chain development without community development is part of the problem the organization has now with its lack of community relevance and recognition. I challenge anyone reading this to walk down any urban street in America and randomly ask someone do they know who or what the NMSDC does, and their answer would be a resounding "no." We cannot have long-term minority corporate supply chain development without community development.

The regional councils would allocate a percentage of their annual revenues to support the national office instead of the current system of the NMSDC supporting the regionals. In this model the pressure is on the national office to provide the national services that make the delivery of services more efficient and more productive.

My proposal was not supported by the NMSDC when I presented this 10 years ago and it is understandable why. (The NMSDC board chose to contract with one of the large corporate consulting firms that produced a costly and beautiful but totally ineffective report and road map.) Power does not like to cede power, even when it is in the interest of the overall organization and its stakeholders. I do not believe what I am calling for represents a reduction in power for the NMSDC. Quite the contrary, this would result in organizational alignment and greater power for the network. The current them-versus-us mentality eventually is unstable, unsustainable, and unproductive.

If the NMSDC made these changes, it would have a chance of surviving and thriving into the next few decades. But I am afraid that without these fundamental structural changes, the NMSDC will not survive. The suggestions here could be the start of a serious conversation about the direction of the NMSDC. That conversation also includes defining what types of leaders the organization needs at the local and national offices.

In closing, I want to say without equivocation that I make these recommendations not as someone who is looking for a job. I am too old for this work, but the role of the gray hairs is to provide the wisdom that only comes with age. I hope that wisdom is reflected here.

CHAPTER 3

Macroeconomic and Microeconomic Policy and Issues

Abstract Chapter 3 covers a range of microeconomic and macroeconomic topics. The global economy has been transformed in recent decades as a result of political, environmental, and technological forces. This chapter looks at how these changes impact labor markets, capital markets and how these changes impact the creation and distribution of wealth.

Keywords Employment · Unemployment · Inflation · Wealth · Taxes · Fiscal Policy · Monetary Policy · Environment

3.1 The Future Is in Leisure

December 9, 2019

In 2015, candidate Donald Trump promised to bring back manufacturing jobs to the U.S. economy. Since January 2017, manufacturing jobs have increased from 12.3 million to 12.9 million in November 2017, growth, but negligible, particularly considering total U.S. employment has increased from 145.7 million to 157.3 million over that same period. Manufacturing job growth represents 6 percent of all new jobs. The overwhelming majority of job growth is taking place outside of manufacturing, and this is likely to continue to be the case in coming years.

F. McKinney, *Critical Issues Facing America*,
https://doi.org/10.1007/978-3-031-93217-5_3

The economic forces driving the distribution of jobs across sectors of the economy are determined by global economic and technological forces that far outweigh the desires of politicians. Those global and technological forces also point to where future growth is likely to occur, and thus where diverse business owners might be well advised to consider in their own company's strategic plans. For the most part, opportunities for diverse businesses will not be in manufacturing.

At the risk of extrapolating from a sample of one, I often think about where markets and technology are going based on my own narrow needs, experiences, interests, and demands. Although I am not a typical American consumer, through observations and data, it appears that American consumers/workers are increasingly going to have more time on their hands, partly because of technology and partly because of rising standards of living. This is not a new phenomenon.

Technology and innovation in production was the key determinant of the Industrial Revolution. Production of clothing in the eighteenth century went from small scale home-based production to large-scale factory production in the nineteenth century. The introduction of lower-cost sources of energy from water-powered mills to coal to oil further increased worker productivity. In 1832, the typical American factory worker worked 68 hours per week. By 1880, the work week had declined to just over 10 hours a day and six days a week. In 1886, the American Federation of Labor announced its support for the 8-hour day. After great struggle, in 1914, Ford Motor company instituted a $5 a day wage and an 8-hour workday. Interestingly, labor productivity at Ford increased more than enough to compensate them for the reduced number of hours. We have been stuck with the 8-hour day for over 100 years. This is about to change.

U.S. workers in all industries work about as many hours as our British counterparts—1680 hours per year. U.S. manufacturing workers averaged a whopping 2,065 hours per year. German workers clocked an average of 1363 hours and experienced a 27 percent higher productivity compared to their British counterparts. Less is more. But before you put in that application to work for Porsche in Zuffenhausen, it is only a matter of time before the American work week also becomes shorter.

A shorter work week will have significant implications on what we do with that time, what economists call leisure. My slightly cloudy crystal ball tells me that the future for business is in capturing a greater share of consumers' leisure budgets. We already see this in the explosion of companies like Netflix to entertain us at home and on our mobile phones. We see it in the gig economy where workers who have more leisure choose to turn their personal cars into transportation services. We see our homes being turned into vacation rentals as traditional hotels struggle to keep up with the growing global demand. Domestic and International air travel in the United States has increased from 700 million passengers in 2003 to over 1 billion passengers in 2018. None of this is possible without increased leisure time, increased productivity, and increased income.

My crystal ball fails miserably when it comes to drilling down to the specific types of technologies and companies that diverse entrepreneurs should think about to take advantage of this massive social change. However, I would suggest that diverse entrepreneurs take a deep dive at how younger consumers and younger professionals are spending their leisure dollars, and at the other end of the age spectrum, how older consumers are spending their time and wealth. The former group sets trends and adopts changes early, while older consumers have the money and wealth to spend unlike their younger children and grandchildren.

Forget about manufacturing. The future is in leisure.

3.2 Dow 30 K and 54 Million Food Insecure

November 28, 2020

President Trump came out of his bunker this week to pardon a couple of turkeys and take credit for the Dow Jones Industrial Average (DJIA) cracking the 30,000-point mark for the first time in history. In normal times, this would be a notable accomplishment. However, we are not in normal times. The DJIA rise must be put in the context of another statistic that came out this week from Feeding America reporting that 54 million Americans are food insecure. In other words, while the stock market is reaching new highs, millions of Americans, including tens of millions of children, are going to bed hungry.

As an economist and someone who once worked for as a securities broker, I know the DJIA is a flawed measure of society's well-being. The DJIA does give us one indicator of how well equity-holding Americans are doing. And while 55 percent of American adults are invested in the stock market, the distribution of Americans who own the stock market is far less egalitarian. The wealthiest 10 percent of all Americans own 84 percent of stocks. Also, there are differences in racial ownership. Sixty-one percent of White households own stocks compared to 31 percent of non-Hispanic Blacks and 28 percent of Hispanics.

This seeming incongruence between the DJIA's performance and the millions of Americans lining up in food lines during this pandemic can be explained by the two different realities facing Americans.

One reality is that of wealthy and high-income Americans who can continue to work, often from their homes, and are experiencing increased savings because there is literally nowhere to go and little to do. In April, the U.S. personal savings rate skyrocketed to over 33 percent. It has since come down to 13.6 percent, but this is significantly higher than the average personal savings rate of 6.5 percent since January 2000. Much of that savings has gone into the stock market, explaining the rise in the DJIA. As more money goes into the market, propping up the DJIA, the wealthy are getting wealthier.

The other reality is that working-class Americans, who had very little savings at the start of the pandemic, have exhausted their savings, millions have lost their jobs, they have lost the $600 a week unemployment supplement provided earlier in the year, they have lost their employer-based health insurance, and now they are on the verge of losing their protection from being thrown out onto the street by their landlords who no longer have to abide by the federal moratorium on evictions. And on top of all this, they and their children are hungry, and they line up for food in the wealthiest country in the history of the world.

President-Elect Biden is going to inherit an economic and health care mess in January that did not have to happen. Instead of reviewing the painful past, there are some things that are clearly going to have to be addressed early in the administration.

First, the administration is going to have to impose strict national restrictions for at least eight weeks in order to starve the virus. Without bringing down the infection rate and the stress on our health care system, the economy, despite the performance of the DJIA, will never recover. Unfortunately, our political system of one president at a time forces us to endure and survive for the last seven weeks of the do-nothing Trump administration. But on January 20, this is job No. 1.

Secondly, the administration, hopefully with Congress' support, needs to reimpose the $600 a week payment to Americans who have lost their jobs and extend eligibility for this benefit until defined economic metrics are attained. This support is necessary to keep people in their homes and keep them and their children fed. This support should be targeted at those who need it most.

Thirdly, the administration needs to redirect the Paycheck Protection Program away from reimbursing small companies to cover their labor costs to a system that helps small companies pay their fixed costs of insurance, utilities, rent, taxes, leases, and bank debt. Absent this reorientation of the PPP, the economic contraction will spread to commercial real estate, banking, insurance, local and state government, and other pillars of the economy. The Biden administration should use the federal unemployment enhancement to take care of workers, thus freeing up resources to support small businesses who also are threatened with eviction and bankruptcy. In a pandemic depression, small businesses do not need help for their workforce when there are no customers.

And finally, the Biden administration should provide unemployed Americans subsidies to purchase health insurance on the Affordable Care Act state exchanges. This pandemic has laid bare the case for moving away from an employer-based health insurance system supplemented by Medicare for the elderly and Medicaid for the poor to a system of health insurance based on citizenship. There is a role for private health insurance companies in this model, but it is unnecessary to have health insurance linked to employment.

I am sure I am not telling Biden anything he does not already know. Americans need help right now. The DJIA could go to 40,000 and it would not feed hungry Americans. Unfortunately, we might be forced to wait until January 20 to make the changes necessary to heal this nation. We literally need to survive until then.

3.3 Biden and the Role of Government

January 22, 2021

One image that will capture the change from the Trump administration to the new Biden administration is the image of President Biden sitting at the Resolute Desk with a two-foot stack of executive orders that were written to undo the executive orders instituted by President Trump.

As I was watching President Biden sign that stack of executive orders reversing the executive orders of his predecessor, I was struck by the enormity of the problems facing the incoming administration and American citizens. Every problem—every executive order can also be thought of as an opportunity to change the direction of America. The list of challenges includes climate change; racial injustice; immigration; voting rights; health equity; public infrastructure; wealth inequality; small and minority business development; student loan debt; homelessness; hunger; women's rights; the opioid crisis; and, of course, the pandemic.

The new administration is boldly trying to address all these complex problems, which have gotten worse over the last four years under the previous administration.

As a society, prior to President Trump, we had two basic camps or economic/political philosophies that addressed society's problems and the role of government. Trump introduced a third camp that included fringe groups like QAnon, neo-Nazis, and the KKK. A recent survey indicated that fully one-third of self-identified Republicans stated they mainly agree with the conspiracy theories of QAnon, and another 23 percent agree with some beliefs of the cult. It is hard to believe that 33 percent of Grand Old Party voters believe Democrats eat children.

Despite the fringes that became mainstream within the Republican Party, their economic beliefs were largely unaffected by the cultists. Republican economic orthodoxy remained for limited government if not outright anti-government. Our recent election was a referendum on whether we choose to solve society's problems by primarily relying on the private sector, or we rely on a mix of the private sector and a functional and competent government.

Former Trump adviser and now-pardoned Steve Bannon pushed Trump and the Republican Party to dismantle what they labeled the "administrative state." The Republican Party viewed the public sector as the "swamp" and the "deep state," something to be destroyed, not celebrated or cherished. The outgoing administration's modus operandi was that markets solve problems and government create problems.

Ronald Reagan famously said in 1964: "This is the issue of this election: Whether we believe in our capacity for self-government or whether we abandon the American Revolution and confess that a little intellectual elite in a far-distant capital can plan our lives for us better than we can plan them ourselves."

The seductiveness of this approach is that the pursuit of personal gain ultimately benefits society at large. In the words of Gordon Gekko in the 1987 movie, Wall Street, "Greed is good." Adam Smith's 1776 economic opus, "The Wealth of Nations" provided the blueprint showing how market forces in a world of small businesses (the butcher, the baker, and the candlestick maker) provided benefits to all consumers. We do not live in Adam Smith's world, yet we continue to be influenced by his views.

In the process of destroying the administrative state, corporate taxes are reduced, and ultimately eliminated. It was Reagan economist Arthur Laffer who came up with the myth that reducing taxes on the rich not only increased economic activity, it increased government revenues, and reduced budget deficits. In this world, the environment is left to the decisions in corporate boardrooms. Wealth and income inequality is dealt with by economic growth, and if for some reason people remain homeless, hungry, poor, uninsured, and economically insecure it is because they are not sufficiently motivated, or they lack the intellectual ability to care for themselves. Republican economic orthodoxy is a one-trick pony—reduce taxes and everything will be OK.

Republican orthodoxy was correct in the sense that the wealthy did become wealthier. It is simply math—reducing taxes on the rich results in the rich keeping more of the earnings that were made possible by a government designed to protect their interests above all other interests.

According to findings by the Pew Research Center, the share of total national wealth going to upper-income families (over $187,000 in annual income) increased from 60 percent in 1983 to 79 percent in 2016. Over the same period, middle-income America's share of total national wealth declined from 32 to 17 percent.

The challenges listed above are related to this issue of income and wealth distribution and the role of government. These things don't just happen. The alternative the Biden administration offers is to allow competent government to address these important tasks. Government is not and cannot be the only solution, but competent government can be part of that solution. A government that believes in its ability address society's problems has a chance to be successful compared to one that believes that government is the problem.

3.4 How to Pay for a $3.5 Trillion Infrastructure Plan?

September 13, 2021

When I was a kid growing up in DC, finding a soda bottle that I could return to the local convenience store for the $0.05 deposit was like winning the lottery. Back in those days, $0.05 could buy you a candy bar. Two decades later when I worked at the Carter White House on the Council of Economic Advisers, I remember my boss, the legendary chairperson of the CEA, Dr. Charles Shultz often quote Senator Everett Dirksen (R-IL) "a billion here and a billion there, and pretty soon you are talking real money." Now we are talking trillions.

Congress is debating a $3.5 trillion infrastructure bill that is difficult to put into perspective or comprehend. With $3.5 trillion you could buy 7 billion round trip tickets to LA from Hartford. With $3.5 trillion, we could buy health insurance for every American family for two years. With this $3.5 trillion the Biden Administration plans to rebuild the infrastructure of the United States. The questions on the table are how we will pay for this and who will pay for it?

Congress is divided on these two questions and Senator Joe Manchin of West Virginia has already come out opposed to the $3.5 trillion plan. This plan, however, is a bargain if the money is spent as planned.

We tend to think of government spending as the same regardless of whether it goes to a Social Security recipient or for infrastructure. President Biden pointed out last week in his address to the nation that we spent over $300 million a day for a total of $1.1 trillion in Afghanistan since we killed Bin Laden. All government spending is not the same. Some, spending, like Social Security, is a transfer from one citizen to another and disappears as consumption spending. Some government spending like military expenditures explodes in death and destruction. Infrastructure spending is more like investment spending. Roads, bridges, airports, and public WIFI systems are consumed over decades. And like all investments, public or private, there should be an expectation of a financial return on that investment.

For too long, we have neglected our infrastructure and as a result we have ended up with an economy that cannot perform at its fullest potential. Any homeowner or business owner knows that if you do not replace the systems, your home or business will decline. We need to make this investment to continue to have the economic growth needed to keep us from each other's throats in a zero-sum game.

The question of how to pay for this investment is inextricably linked to who should pay for this improvement in American infrastructure. And who should pay should be tied to who benefits from the infrastructure investment. First, all Americans should pay something for this investment in the quality of their lives.

According to the Federal Reserve, there is $137 trillion in household wealth in the U.S. economy. Household wealth grows every year because much of it is invested in assets that are growing in value. Most of this wealth is not in the form of cash, or what economists call liquid. However, a 1 percent tax on household wealth, which everyone should pay, would generate enough cash over three years to pay for the $3.5 trillion infrastructure bill.

The logic of paying for this investment with wealth taxes instead of traditional income taxes is overwhelming. Every American will benefit from building a better more sustainable American infrastructure. The wealthy should contribute the most because they will benefit the most.

Democrats in Congress are focusing on raising income taxes on American individuals earning more than $400,000 per year or dual earners making more than $450,000. There is a case to raise income taxes, but income taxes are not the only option and may not be the best option for this effort, given the investment nature of infrastructure. This public investment will benefit all citizens for decades.

According to the Brooking Institution, the typical African American household has approximately $17,000 in household wealth. Their wealth tax of one percent as I propose to finance the infrastructure bill would be $170.00. The typical White household has $170,000 in wealth. Their infrastructure tax bill would be $1,700.00. An American billionaire with $1 billion in net worth would pay $10 million a year for three years.

Sounds unfair until you consider that wealth grows. In 2020, according to the Federal Reserve, household wealth grew by 3.8 percent. Even if it grows at 3 percent, the average African American household will see its wealth grow from $17,000 to $17,510. The billionaire will see her wealth grow by $30 million, more than enough to pay their infrastructure tax bill and still put an additional $20 million in their pocket.

Investing in infrastructure will make it possible for all Americans to become wealthier. We cannot rely on the already wealthy and their representatives in Congress to make these decisions for Americans who are simply trying to build a sustainable economy that will provide for their children, now and in the future.

What about the wealthy who have their wealth in illiquid assets like land or real estate? How will they get the cash to pay their wealth tax? Not to minimize this issue, but this is a financing problem, which is one of the reasons we have banks and financial consultants.

We are a rich country. We must stop thinking like a poor one and stop trying to run our economy on the cheap. We can do this, and it makes sense to make the $3.5 trillion investment and pay for it with a small wealth tax, limited to three years. We will all be better off, and we can all say we paid our fair share.

3.5 Anti-Trust and a Wealth Tax

April 29, 2022

I drive a Tesla. It is the best car I have ever owned. But I do not like the move Elon Musk just made to buy Twitter. My concern and problem with Musk's purchase of Twitter is about concentrated economic power that easily is transformed into anti-democratic political power. The concentration of wealth in America is worse today than it was right before the Sherman Anti-Trust Act was passed in 1890.

The Sherman Anti-Trust Act was a democratic response to the growing economic power of the "robber barons" that emerged after the Civil War when the United States was experiencing an economic transformation driven by new modes of transportation, communication, and consumer products. The men who controlled steel, railroads, finance, oil, and shipping paid no federal income taxes on their vast income and wealth even after the Sherman Act. The federal income tax did not begin until 1913. These men and their closely held companies used their economic power to buy political power controlling state legislators, mayors, congressmen, senators, and even presidents.

The first major cases brought under the Sherman Anti-Trust Act did not occur until the United States v. Standard Oil, Rockefeller's giant oil monopoly in 1911. The case ended up in the Supreme Court which ruled that Standard Oil had restrained trade resulting in higher prices, reduced quantities, and reduced quality. The remedy was to break up Rockefeller's company. Supreme Court Justice John Marshall said the following:

> All who recall the condition of the country in 1890 will remember that there was everywhere, among the people generally, a deep feeling of unrest. The nation had been rid of human slavery, fortunately, as all now feel,— but the conviction was universal that the country was in real danger from another kind of slavery sought to be fastened on the American people; namely, the slavery that would result from aggregations of capital in the hands of a few individuals and corporations controlling, for their own profit and advantage exclusively, the entire business of the country, including the production and sale of the necessaries of life. Such a danger was thought to be then imminent, and all felt that it must be met firmly and by such statutory regulations as would adequately protect the people against oppression

and wrong. Congress, therefore, took up the matter and gave the whole subject the fullest consideration.

The country is now again facing the danger Justice Marshall wrote about over 100 years ago. Fortunately, there is a solution to this problem. That solution is to tax wealth and enforce the Sherman Anti-Trust Act and have smarter federal tax laws.

The federal government should first eliminate the cap on Social Security taxes which now is at $147,000 per year. It makes no sense to limit this tax to pay for the support of America's senior citizens, many of whom live near or below poverty on their Social Security income. While Social Security effectively lifts over 16 million seniors out of poverty, nine percent of Americans over 65 still live below the official poverty line. All these people could be lifted out of poverty with the elimination of the Social Security income limit.

Secondly, our tax rates need to be increased on all income from all sources. Currently, the tax rates on incomes range from 10 percent on incomes under $20,550 for married couples to 37 percent on married couples earning $647,850 or more. We need more brackets. There should be a bracket for $1,000,000 to $10,000,000; from $10,000,000 to $50,000,000; and for over $50,000,000. And those rates should be greater each step along the way. My suggestion is that the minimum tax on incomes over $50,000,000; 50 percent.

We also need to revamp our estate taxes or what critics call, death taxes. Why should wealth so easily be passed from one generation to the next?

Capitalism works best when entrepreneurs and innovators have a chance to develop. Concentrated wealth that is passed down from generation to generation encourages political shenanigans not economic growth. No American should be allowed to pass down more than $50,000,000 to their heirs in total. Any wealth over that should be taxed away and spent on the general welfare of society that allowed for the wealth generation in the first place.

I can hear the screams from Greenwich in my basement in Trumbull. "You are recommending killing the goose that lays the golden eggs." I would only ask that the makers of this argument explain to me and most Americans why $50 million is not enough to leave as a legacy.

If we had a tax system that made economic sense, we could avoid the problem of Elon Musk and his fellow American oligarchs owning key companies in already highly concentrated industries. We know we have a problem when the average American can name the dominant companies in cellular communication, pharmaceuticals, oil and gas, automotive, social media, and computers. The American economy can be even more powerful than it is if we do not allow concentrated wealth, concentrated power, and concentrated industries. The federal government should not allow Musk to buy Twitter, and this industry needs to be broken up, just as Standard Oil of New Jersey was in 1911.

3.6 Don't Cut Gas Taxes—Make Metro North Free

June 22, 2022

A central theme of Governor Lamont's reelection campaign is cutting state gasoline taxes. In ads that have flooded the airwaves, families, businesspeople, women, and drivers all speak about the financial benefit they are expecting because of the Governor's action. Now President Biden has proposed a three-month federal gas tax holiday. Both the state and the federal gas tax holidays are temporary and largely political. Neither makes much sense economically or more importantly environmentally. A far superior policy would be to maintain the taxes on gasoline consumption, I could even argue they should be increased but I will not for fear of my life or being kicked out of the Democratic Party. The superior policy that would put more money in the hands of Connecticut households, reduce gas consumption and prices, and contribute to a cleaner environment is to make Metro North ridership within the State free.

To the Governor's credit, some bus rides are now free. Extending this to the trains would have a significant impact on getting the cars off I-95. We should also prioritize expanding the rail lines between New Haven and Springfield, MA, with a stop at Bradley. The latter would have a similar salutary impact on I-91 traffic.

According to the U.S. Department of Energy, non-commercial, non-industrial gasoline consumption (i.e., consumer gasoline consumption) in the state is about 1.2 billion gallons a year. We must remind ourselves, that the source of gasoline is non-renewable, and it transforms into harmful climate killing gases. At a time when forests are on fire, hurricanes, and other violent storms are normal not rare, and sea levels are rising threatening coasts and food supplies, now is not the time to make gasoline consumption easier.

The six-month state gas tax holiday will cost the state about $150 million. Economists know that this does not mean that Connecticut residents will save $150 million. It is possible, even likely, Connecticut consumers could pay more for gas at the pump during the gas tax holiday than they do now. Taxes are only a small part of what determines the price of gas. How do you think consumers and taxpayers are going to feel if gas prices do not come down by the $0.25 charged by Connecticut and the $0.18 charged by the Federal government?

The wiser long-term strategy would be for the state to make Metro North train travel within the state free and pay for this with gas tax revenues. Getting people out of their cars needs to be the priority right now.

This policy addresses both the personal costs of inflation on households and the responsibility of those households to do their part to improve the environment. Making the trains free helps solve the problem of inter-city gas consumption but admittedly does little to reduce intra-city gasoline consumption. Getting people out of their cars in their commute to work could have significant impacts on demand and ultimately the price of gas.

The age of the gas-powered automobile is ending. State and federal policy should accelerate that transformation, not slow it down. The policy to temporarily suspend gas taxes is nothing more than a gimmick that will not relieve the problem American households are experiencing because of forces beyond their control, and beyond the control of the federal government. To prove this point that inflation is global and beyond politics, is the fact that inflation in the Netherlands is 10.2%, in the UK it is over 9 percent, in Germany it is 8.7%, and in Italy it is 7.3. Inflation in the United States is now over 8%, which is not unlike what is taking place in other Western economies. Market forces determine gas prices. International comparisons prove the inflation we are experiencing is not about Biden.

We and other countries around the world are experiencing inflation because we all experienced a global pandemic and all our governments attempted to reduce the impact of that pandemic by pumping out money and programs to support households and consumer to prevent the world and their economies from experiencing something even worse—a great depression.

This does not mean we should sit idly by while households suffer. The Biden and Lamont Administrations' reductions in gas taxes while well-meaning, will not work to ease the pressure on American households.

There are only two ways to reduce inflation and gas prices; one is for demand for gas and other goods and services to decline, and the other is for the supply of gas and other goods and services to increase. Currently, we are experiencing frustrated consumers who want all the things they could get pre-pandemic—right now—while at the same time the pandemic's destruction of global supply chains reduced the supply of just about everything. This perfect storm of high demand and low supply is the reason prices of all goods are rising in this country and around the world. The war in Ukraine is a factor, but inflation pressure predates the war. As painful as this is to hear, inflation is part of the correcting mechanism of global capitalism. Rising prices and rising corporate profits will lead to greater supply as greed replaces fear in the minds of capitalists. And rising prices will begin to reduce consumers' demand for those same goods and services. We should not forget that gas prices averaged $2.17 in June of 2020 because there was no demand for gas during the lockdown in the early phases of the pandemic.

Making Metro North train rides free in the state reduces the demand for gasoline, speeds up the necessary adjustments, and is good for the environment. The state and federal government should take the resources gas taxes provide and invest heavily in public transportation and the electrical vehicle infrastructure.

The current policy of reducing gas taxes is being done under the guise of an anti-inflation policy, but it really has nothing to do with combating inflation and will be ineffective. Let's not let this inflationary period blind us from what really must be done to reduce gas prices and the inflationary burdens on Connecticut consumers.

3.7 The Consumer Price Index Is Not Your Cost of Living

July 22, 2022

The Consumer Price Index (CPI) is front page news when it is released, but it is clear to me that many Americans do not know the difference between inflation, the CPI, and their personal cost of living.

The most recent CPI reported an annualized increase of 9.1 percent, the highest increase in forty years. As an economist who worked on various aspects of the CPI while on the Council of Economic Advisers during the Carter Administration, I know that most Americans do not know how the CPI is constructed or computed and therefore do not know that the CPI most likely does not reflect their cost of living, nor is the CPI the most comprehensive measure of inflation in the economy. This lack of understanding can lead to unnecessary panic and bad economic policy by political and business decision makers who should know these distinctions.

Before addressing some of the problems with the CPI inaccurately reflecting your cost of living, I must state clearly that inflation, particularly high and increasing rates of inflation imposes costs on consumers, investors, small businesses, governments, and others. You might have seen those grainy images of Germans after World War I carting wheel barrels full of near valueless German marks to the store to buy a loaf of bread. This is extreme hyperinflation that will destroy any society, government, and economy.

Hyperinflation is not what we are experiencing today. Moderate rates of inflation, like what we are experiencing can be managed, but historical experience suggests this will not be without a cost, and that cost is a recession. But it is the image of increasing rates of inflation that is driving the fear of inflation. Shedding light on the CPI and why it is not a cost of living index for most consumers is not to downplay the costs of moderate inflation, but to reduce the political hysteria on the subject.

The CPI is reported at an annualized rate. This means that the 9.1 percent increase is for the period, June 2021 to June 2022. The CPI or the rate of inflation did not increase by 9.1 percent from May 2022 to June 2022. The actual increase in the CPI in June was 1.3 percent.

In capitalist economies, it is common for some prices to go up and others to go down because of fluctuations in demand and supply for goods and services. For example, gas prices rose in June primarily because of changes in demand and supply of gasoline. In June, increases in gas prices accounted for more than half of the total increase in the CPI. Gas prices have been falling recently. If nothing else changes, the July CPI will be lower than the June CPI.

The CPI is not a measure of your cost of living because of how the CPI is calculated. To calculate the CPI, the Bureau of Labor Statistics creates a market basket of goods and services for consumers. This market basket is updated regularly but is designed to represent what on average urban consumers buy. The market basket includes housing, energy, food, apparel, electronic goods, travel, health care, and other goods and services consumers' purchase. Housing is the largest component of the CPI representing over 32.3% percent of the index followed by food (13.4%), energy (8.7%), Medical care services (6.8%), and transportation (5.9%), and everything else at 33%. These weights are based on the experiences of actual consumers, but chances are, your weights are different. If you own your own home, your housing expenses are likely significantly below the official CPI weight for housing. If you own an electric vehicle or do not have one at all, your transportation expenses might be lower. If you own a Hummer (the non-electric version) your transportation and energy weights will be higher. If you are healthy, increases in your cost of living are lower.

The CPI is helpful in capturing the direction of the general level of prices and helps decision makers make important decisions on wages and the prices of goods under their control. The CPI impacts workers who must make calculations on whether the job being offered is at a high enough wage to cover their expenses. The CPI or other price indexes are used to adjust wages, benefits, and prices. The CPI is better than flying blind.

Economics often involves tough choices. We have been spoiled by 12 years of economic growth and stable prices. In this current climate you can ask: Would you rather have a higher probability of being unemployed and pay less for goods and services, or would you rather have a job and pay more? Rapid increases in prices impose costs on consumers, investors, lenders, and governments. But it is important to keep things in perspective. I hope the Federal Reserve does not overreact and push the economy into a recession to squelch inflation.

3.8 ESG, Capitalism, and Government Responsibility

March 4, 2023

"Sustainable investing is the practice of analyzing a company's environmental, social and governance (ESG) risks, as well as assessing its opportunities and progress, using ESG data and fundamental insights, to inform the allocation of capital." Black Rock.

If you are confused about ESG, you are not alone. This week, Washington and Wall Street are in a tizzy because of a Biden Labor Department regulation that would allow investment companies to consider Environment, Social, and Governance (ESG) in retirement account investment decisions. It is not a surprise the House and its Republican majority would vote to block the regulation. And given the upcoming 2024 Senate elections that some vulnerable Democrats in red states might use this as an opportunity to distinguish themselves from the Administration and other Democrats. The vote by both houses to block the rule now ends up on the President's desk where Biden could use his first veto. If Biden does veto the bill, the Senate lacks the votes to overturn the veto. But this is about the most wasteful exercise of political nonsense in recent memory, and it was completely avoidable and unnecessary.

The regulation does not require investment advisors to consider ESG, it just says they can. Investment advisors do not need an act of the President or an act of Congress to consider all factors in making allocation decisions. ESG has gotten a bad rap as "woke" capitalism. Investors have always considered factors other than profitability in their investment decisions. Recently, for example, a meat processing company has been accused of hiring underage migrant workers to clean dangerous slaughterhouse floors and equipment. The accused company may generate higher profits because it hires migrant children who work the midnight shift (after attending a full day of school) at below minimum wage rates. If investors knew of these illegal practices, this particular company's profits and attractiveness will be diminished because of the legitimate fear that the offending company would be revealed for being a child labor abuser which would lead to lower consumer demand, fines, penalties, and lawsuits, all resulting in falling stock prices.

Whether it is child labor law violations, chemical spills, discrimination, or corporate malfeasance of any type, ESG is designed to encourage companies to do the right thing in their pursuit of profit. A mini-industry has developed, overnight, to help companies measure and manage ESG for corporate boards, employees, consumers, regulators, investors, and other stakeholders.

ESG is not a new corporate phenomenon. Some point to former United Nations General Secretary Kofi Annan's push for global corporations to take more responsible roles to curb climate change. There was an unquestionable boost in ESG after the George Floyd murder in 2020. Many companies after the racial justice protests made very public announcements regarding their commitments to ESG.

The vote by both houses of Congress to overturn the DOL regulation allowing public investment companies to use ESG in their allocation decisions now goes to the President's desk. If Biden signs the bill, he defeats his own administration. If he vetoes the bill, it sends a message to his base that ESG is a legitimate factor in investment decision making.

The political problem masks the bigger question that has not been addressed about ESG. Whose responsibility is it to ensure that companies operate safely, humanely, and follow the laws and interests of all consumers, workers, and taxpayers? My answer to that question is that it is the responsibility of government to establish the rules of the game in capitalist society and for corporations who operate under those laws. ESG is a poor substitute for appropriate federal (and state) government oversight of corporate behavior. My position against ESG is different from the position of Republicans opposed to ESG. Republicans oppose ESG because they do not want any restrictions on the "rights" of companies to do as they please, whether self-imposed or by government. I am opposed to ESG because it is a poor substitute for legitimate robust government intervention and control of corporate behavior.

If I hired underage migrant children to clean my house and shovel my snow, I might get arrested. Corporations who violate the law often get away with a simple slap on the wrist. Norfolk Southern, the train company that had the "accident" in Ohio will pay a $1.5 million fine. That level of fine is no disincentive to do the wrong thing—it's just a cost of doing business.

Until the federal government takes its regulatory role seriously and uses the big stick that it has, corporations will attempt to self-regulate through efforts like ESG, and will do it poorly. Accidents do happen and should be distinguished from purposeful bad acts, but the federal government must clearly and firmly establish the rules of corporate behavior. Citizens have the legitimate right to expect corporations to behave and when they do not behave, there should be grave consequences. The Supreme Court in the Citizens United case established, at least for now, that corporations have rights like those of natural citizens. Citizens United was a terrible ruling, but corporations cannot have it both ways. If the federal government did its job, we would not need ESG.

3.9 Do We Need Banks?

March 30, 2023

The financial crisis of 1907 started when some Wall Street schemers borrowed outlandish sums from the Knickerbocker Bank to try to corner and control the copper market. Their plan was to control the price of copper, the fiber optic cable of the second Industrial Revolution. Unfortunately for them, they failed; and with their failure, panicked Knickerbocker depositors literally ran to the bank to withdraw their funds. As is always the case with all banks in all countries, banks rarely survive a run.

The Silicon Valley Bank's ("SVB") recent failure was just a modern version of an age-old story about the fragility of banks in the face of a crisis. Due to the actions of the Federal Reserve and the largest private financial institutions in the world, the seeming resolution and containment of the crisis regarding two other financial institutions was swift and decisive. Citibank, Wells Fargo, JP Morgan Chase, Bank of America, and others came together to make real the Federal Reserve's promise that no depositor would lose a penny and no taxpayer dollars would be used in this resolution, even if their deposits exceeded the $250,000.00 limit on FDIC deposit coverage. Apparently, part of the problem at SVB was that too much of their deposits were held by depositors with accounts far more than the FDIC limit, making those deposits vulnerable to being wiped out at the slightest hint of run.

From 1999 to 2017, I was an economic consultant to the Federal Home Loan Bank of San Francisco ("FHLBSF"). The Federal Home Loan Bank system was created after the Great Depression to stabilize savings and loan institutions ("S&Ls") by assisting their member institutions with access to credit. The FHLB also regulated the banks and collected information on consumer mortgage loans, called loan application records ("LARs"). Every application for a mortgage to every financial institution is recorded. My work was to review the millions of LARs to help the FHLPSF identify problems in the market. Our work was important because in the late 1980s and early 1990s, the S&L industry had gone off the rails when S&Ls were deregulated by the Reagan Administration and Wall Street schemers stepped in and bankrupted over a third of all S&Ls in the country.

I fully understand the economics of not letting one bank's failure become a contagion impacting hundreds if not thousands of other banks because of the bad actions of one set of bank executives.

It was financial crises that led to the 1907 depression, the Great Depression in the 1930s, the Savings and Loan crisis of the late 1980s and early 1990s, and the Great Recession in 2008–2010. All these crises can be traced to the lack of good regulatory oversight and injurious behavior by financial schemers.

Consumers, workers, firms, and governments need a place to park their paychecks, pay their bills, collect other payments, and borrow money. The economy needs to be able to connect those with resources to those who can put those resources to work while making money for both parties.

But banks are by design—unstable. Depositors put their money in the bank. The bank takes some of those deposits and lends them to borrowers, leaving just enough cash in the bank to meet the anticipated cash needs of the bank's depositors. Economists call this intermediation.

The first banks started out simply as places to store your gold, diamonds, and cash. The banks realized their customers did not come in every day to check the security of their assets. These early financial geniuses realized they could make more money using other people's money by lending assets they did not own to borrowers who promised to bring the money back with interest. Banking is a great business model when the bankers do not get too greedy.

Our economy because of technology is becoming "cashless." We do not use gold for currency, and we are increasingly using digital transactions. We get paid without the need to go to the bank to deposit our checks; those funds go directly from our employer's bank to our bank. A simple transaction on a digital balance sheet. We use credit and debit cards more than cash to execute our daily transactions. We pay our babysitters by sending funds through our cell phones.

Additionally, recent data from the Federal Reserve shows that the median amount of cash Americans have in banks is $5000, and the bottom forty percent (in terms of income) have accounts of only $2000. Clearly the banks for most Americans are used primarily as transaction vehicles and not places to store valuables or to generate wealth.

While Americans are not great savers like the Japanese or the Koreans, we are great borrowers. According to a 2023 study by NerdWallet, the typical American household owes more than $165,000. Access to credit allows households to borrow money for higher education (student loan debt), purchasing homes (mortgage debt), and just buying stuff (credit card debt). Banks are involved in all this debt.

Are we getting to the point in history, society, and culture when banks and traditional capital institutions are solutions to the problems of the past and not the opportunities of the future? Is a bank-less economy possible? This may sound like throwing out the baby with the bath water. Having periodic financial crises may not be a reason to contemplate a bank-less world. However, in a bank-less world, worth and value will physically reside within the person not in a building.

Something or someone will always have to keep account of what is coming in and what is going out. Right now, our complex financial system manages this process. This future sounds like the promises of digital currency. My view is bitcoin, and their class of assets are a transition not the end state.

To some, a bank-less world may sound like chaos and anarchy. There is no disputing the fact that we are experiencing changes in financial technology that were unimaginable 40 years ago. The only thing that is certain is that 40 years from now, banks will be unrecognizable.

We need banks today, but I do not think the banks of the future will be needed in the way they were needed in the past. A bank-less economy may not mean the future will be free of financial crises. Whatever replaces these ancient institutions will require robust regulatory oversight. Schemers with access to other people's money and unregulated greed will always result in financial instability.

3.10 We Have a Wealth Problem

June 29, 2024

In my last column, "What is the True Nature of the Trump Coalition," I discussed the alliance between billionaires and the MAGA supporters who are about as economically far away from each other as the North Pole is from the South Pole. Despite the significant material differences in how they live their daily lives and what they want from society, not just government, they are united in their support for the convicted felon former president whose policies are simply to give the wealthy even more control and keep the MAGA crowd entertained.

Soon after my last column, we learned that an heir to the Mellon family wealth, Timothy Mellon, contributed $50 million to Trump and $25 million to RJK, Jr. I can only speculate that the RFK, Jr. contribution is a financial hedge, providing the support RFK would need to stay in the race (under the theory that the more people in the race, the better it is for Trump). If Mellon believes this, I would not be surprised if he and his billionaire friends started writing checks to Jill Stein and Cornell West.

This week, we learned that pro-Israel billionaire heir to the Adelson Las Vegas casino fortune, Miriam Adelson, is considering a $100 million contribution to the Trump campaign. One thing is clear from the past two weeks, Trump will get all the support he needs from the billionaire class, and he will happily do their bidding on income taxes, wealth taxes, estate taxes, corporate taxes, health and safety regulations, labor regulations, environmental regulations, Israel, and anything else they want. If Trump is elected, it will be the result of him selling the country to the billionaires. I do not claim my column caused any of these actions, but it was accurate in its prediction on the billionaire class's support for Trump.

My last column also discussed what MAGA wants and why they are part of this unholy coalition. We saw their goals and objectives demonstrated in Louisiana last week. MAGA state legislators passed a law requiring the posting of the Ten Commandments in every public school in the state, contravening the constitutional separation of church and state. It might be helpful to remind Americans what Thomas Jefferson had to say on this topic in a letter to the Danbury (CT) Baptist Association in 1802.

> Believing with you that religion is a matter which lies solely between Man & his God, that he owes account to none other for his faith or his worship, that the legitimate powers of government reach actions only, & not opinions, I contemplate with sovereign reverence that act of the whole American people which declared that their legislature should "make no law respecting an establishment of religion, or prohibiting the free exercise thereof," thus building a wall of separation between Church & State.

Conservatives have often relied on "originalism" in their understanding of the intentions of the founding fathers. Jefferson's statement above makes clear—and leaves no room for interpretation—that what the MAGA crowd is attempting in Louisiana is unconstitutional and will be struck down. There is no doubt what the founders intended when it came to this matter.

The MAGA Christian extremists do not care about the Constitution; nor do they care about Supreme Court precedents. They are counting on Trump appointed and other conservative judges and justices to use sophistry to interpret the law in their favor. They only care about their perverse interpretation of religion.

The billionaires are not motivated by these religious issues. They could not care less. I do not think they pay much attention to Matthew 19:24: "It is easier for a camel to go through the eye of a needle than for a rich man to enter the kingdom of God." The billionaires are happy to align with MAGA supporters on these religious issues so long as the conservative judiciary aligns with their interests. Most importantly, billionaires care about more tangible things, in the here and now. For them, the most tangible of all things is their money.

If the United States does not seriously address the accumulation of huge fortunes that last centuries, democracy is doomed. One person one vote will not work where one voter gives $100 million to a candidate.

Archibald Mellon was born in 1816 in Northern Ireland. His grandson, Thomas Mellon, founded the Mellon Bank in 1869. Thomas's son Andrew Mellon became the principal investor in Gulf Oil and Alcoa. The Mellon family also owned or continues to own major stakes in The Pittsburgh Tribune-Review, Koppers, New York Shipbuilding, Carborundum Corporation, Westinghouse Electric, H.J. Heinz, Newsweek, U.S. Steel, Credit Suisse, First Boston, and General Motors. Their original bank is now BNY Mellon.

The Mellon family is not alone in their influence. Think about the damage Elon Musk can do with the $40 billion bonus his board just approved! That is enough to give every American $100 and he still would have $10 billion left. At a minimum, we need immediate legislation that restricts private political contributions, and we need to totally restrict corporate contributions of any kind to political candidates or elected officials. And yes, we must have a serious conversation about our tax system that allows Americans who are eight or nine generations away from the source of the creation of their wealth to continue to have an outsized influence on the politics and economics of this country. No American should be able to write a check and determine policies that will impact all of us like these billionaires are doing today.

3.11 Biden Economics Worked but It Is Time (Published the Day Before President Biden Withdrew from the 2024 Presidential Race)

July 18, 2024

In 1936, John Maynard Keynes' book the ***General Theory of Employment, Interest and Money*** was published. It was during the depths of the Great Depression that the ideas in that classic book described how the economies of Western Europe and America had descended into a prolonged decline in economic activity. Unemployment rates of over 30 percent, bank failures, stock market crashes, soup lines and the near destruction of Western democracy was the nature of those times. Keynes' solution was to put aside the traditional economic policies being promoted, namely that workers should accept lower wages to make them more attractive to employers. Keynes showed that this and other free market policies failed to understand what we now call macroeconomics. Keynes proffered that the only way to get the economy back on track was for the federal government to spend money that it did not have, i.e., deficit spending. At a time when balanced budgets were the dominant belief, his solution was economic heresy. But it worked.

Fast forward to 2020. On January 22, in response to a question about whether he was concerned about COVID, President Trump said, "No. Not at all. And we have it totally under control. It is one person coming from China, and we have it under control. It is going to be fine." As late as August 5, Trump stated, "It is going away. No question in my mind that it will go away, hopefully sooner rather than later." We were in the midst of a full-blown pandemic. And while, more Americans died from COVID during the Biden Administration, the problem started with Trump's inaction and dissemination of false and misleading information which caused confusion, understated the magnitude of the situation and wreaked havoc.

Notably, Trump's denial of COVID-19 pandemic contributed to him losing the 2020 election. America wanted and needed new leadership.

Instead of the laissez-faire attitude of Trump, the incoming Biden Administration took a hyper activist robust federal approach reminiscent of the Keynesian approach to the Great Depression. Fundamentally that approach focuses on the here and now.

President Biden signed the $1.9 trillion American Rescue Plan on March 11, 2021, less than two months after taking office. This landmark legislation put cash into the hands of Americans earning less than $80,000 per year. That money was necessary to replace the lost earnings of workers who had been furloughed. The bill also extended unemployment insurance payments to $300 per week, and temporarily extended the child tax credit, increasing the amount to $3000 per child for children between ages 6 and 17 and $3600 for children under 6. Importantly, the child tax credit resulted in the single largest reduction in childhood poverty in the nation's history. The bill also enacted the $7.25 billion Paycheck Protection Program which supported small businesses and nonprofit organizations in continuing to employ their workers during the time the economy was effectively shut down. The bill invested over $128 billion in local school districts to support towns and cities in red and blue states. The bill also provided for low-income Americans, restaurants, and renters.

We seem to have forgotten that these programs prevented the U.S. economy from slipping into a prolonged and catastrophic depression. It is a human trait that we do not take heed to the warnings of disasters that we narrowly escape.

The result of this bill and other Biden Administration executive orders and legislation, the Inflation Reduction Act in particular, along with the very real supply chain problems was the reason we experienced the inflation of recent years. The Federal Reserve did its job in the face of these massive Keynesian expansionary policies by increasing interest rates knowing that inflation would be a consequence of the Biden Administration's actions. Biden economics worked.

While there are differences of opinion among professional economists on whether President Biden went too far in his expansionary policies, there is no respectable economist who believes that if no action had been taken, the U.S. economy would have recovered as quickly as it did.

President Biden's management of this crisis was masterful. His emphasis on building the recovery from the middle out instead of from the top down demonstrated clearly that President Trump's trickle-down economics with its corporate tax cuts and tax cuts for billionaires and deregulation of industries, was not only not necessary but damaging to the economy and society.

There is unrelenting focus now on President Biden's age. But Biden's age is an unfixable problem in the face of a surging Trump campaign. It is with great sadness that I find myself in favor of continuing the Biden approach to our economic and social problems without President Biden at the top of the ticket. His legacy has a better chance of enduring only if Democrats control the White House and both Houses of Congress. That should be the goal. We will see what happens on Election Day. But whatever happens, we all owe President Biden a thank you.

CHAPTER 4

Race, Politics, and Economics

Abstract This chapter on Race, Politics, and Economics is central to the issues facing Black Americans. All Americans other than the Native Americans are here as the result of voluntary immigration to this country. Black Americans were brought here against their will, and laws from the founding of the country were designed to maintain their second class status. The legacy of slavery still plays a large role in current policies and circumstances. This chapter reviews those powerful dynamics.

Keywords Race · Politics · Economics · Blacks · History · Slavery · Discrimination · Racism

4.1 When Anti-Asian Racism Hits Home

March 19, 2021

The fall semester in 2018 was my first semester as a professor at Quinnipiac University. Earlier that semester, I met with student government leaders to introduce myself as the director of the People's United Center for Innovation and Entrepreneurship. One of the student leaders was a Chinese national international business major who was active on campus and offered to volunteer to help the new center. I welcomed her help.

F. McKinney, *Critical Issues Facing America*,
https://doi.org/10.1007/978-3-031-93217-5_4

She helped by marketing our activities on campus and arranging for me to meet with other student groups on campus.

One day near the end of the semester, I offered my friend, whose name I will not use, a ride to her dorm at Hilltop. On the way up the hill, I asked what she was doing for the holiday break. She informed me she would be staying in her dorm for the month.

After saying goodbye, I went home and informed my wife that a Chinese national was staying in a near-empty dorm during the holiday, with no car and little access to food. We were about to have our extended family over for Christmas. Our holiday celebration included my son, my daughter-in-law (who was born in Iran), my grandson, my daughter, and two of my daughter's graduate school classmates. One of my daughter's classmates was a Korean national and the other was a Taiwanese national; neither of my daughter's classmates had anywhere to go during the break. This was the second year of this international gathering where we broke bread together and played music. I told my wife I would like to invite my new Quinnipiac University friend to dinner.

Christmas came and my student friend took Metro North to Bridgeport. I picked her up and once again we had an enjoyable Christmas dinner with plenty of stories about our lives. When it was time for my friend to go, I knew from the look in her eyes that she did not want to go. So, we invited her to stay over. She stayed for six weeks.

We had a marvelous time. We took her to New York City to the Metropolitan Museum of Art. When we visited the Chinese section in the museum, she gave us insights into the art and the history of China. We were extremely impressed with the detail of her knowledge of Chinese art and history. At first, I could not believe what she was telling us was accurate. But after hearing her description, and then reading the description near the art, it turned out she was 100 percent correct.

We took her to our favorite restaurants. One of those restaurants was our favorite Chinese restaurant in Westport, one my wife and I had been to many times. The owner came by to take our order, and asked our Chinese student in Cantonese if she spoke the language. They struck up a conversation in Cantonese that neither my wife nor I could understand. When the owner left with our order, our student told us that the restaurateur wanted to know what her relationship was to this black couple. Our Chinese friend told her that my wife and I were her parents. This is what led to the extended conversation between our student and the restaurant owner. This demanded an explanation.

From that day on, our Chinese student referred to us as "Mom and Dad." We lovingly accepted our new child and our new roles.

I tell this story to say that my Chinese daughter is afraid to walk the streets of Connecticut for fear that she will be randomly attacked. As her adopted father, I am also afraid for her safety and all our Asian brothers and sisters. This fear is not without merit. My daughter has been verbally assaulted during this pandemic on numerous occasions by complete strangers. For those who do not think this is real—it is real.

What can we do about it? In one of my columns last year, I said that America's virus is racism. While not unique to America—unfortunately, there is racism in far too many places—racism is a sickness. Racism must be fought—it cannot be allowed to fester untreated. Children of all races must be taught to look beyond color and see the humanity in all of us. Some adults are beyond repair, but racists must feel that their behavior, their attitudes, and their comments are not acceptable, and racist actions will be punished by the full extent of the law.

People of goodwill must stand in solidarity with our Asian brothers and sisters and say these attacks will not be tolerated. Martin Luther King famously stated that: "Injustice anywhere is a threat to justice everywhere."

Speak up if you see Asians being verbally abused or threatened. And more importantly, if we are to fight racism, everyone must try to get to really know people who are racially and ethnically different from ourselves. When we do this, we will recognize our common humanity.

4.2 Trump Promises to Black Americans Ring Hollow

October 17, 2020

President Trump likes to say he is the best president in history for Black Americans except for Abraham Lincoln. As evidence, he points to the historically low Black unemployment rate prior to the pandemic. He also takes credit for rising incomes and reduced poverty rates among Black Americans, also pre-pandemic. The pandemic and the economic consequences of the pandemic will result in Black Americans being far worse off at the end of his administration than at the beginning.

While the pre-pandemic facts are on his side, his explanation for these positive economic factors are debatable. It is also a fact that these numbers were declining when he took office in 2017 during the Obama/Trump Expansion from 2010 to 2020. Separating out whether the positive economic statistics are attributable to Obama or Trump, or both is something economic historians will have to decipher.

It is also a fact that President Trump is frustrated by the lack of support from Black voters given his self-perceived success in helping Blacks. Recent polls give Joe Biden a 75-percentage point lead over Trump among Black voters.

Before the 2016 election, Trump asked Black Americans, "What do you have to lose?" In his mind, he delivered, and had it not been for what he calls the "China Virus," his value to Black Americans would be plain for all to see and only the most die-hard anti-Trumpers would not be solidly in his corner.

Now, just weeks before the election, President Trump has rolled out the Platinum Plan for Black Americans. The plan promises to make Juneteenth a national holiday. The plan also would designate the KKK and Antifa as terrorist organizations.

On the economic front, the plan promises to invest over $500 billion in capital and provide significantly more federal contract opportunities to support Black businesses, create 3 million new jobs for Black workers, invest hundreds of millions of dollars in Black-owned banks, and in historically Black colleges and universities. Wow!

I would have to be a fool to say that these are policies that are not good for the Black community and America in general. These are plans Biden should consider. The problems are timing and history.

I am reminded of the Stevie Wonder song, "You Haven't Done Nothin'":

Why do you keep on making us hear your song?
Telling us how you are changing right from wrong
Cause if you really want to hear our views
You haven't done nothing.

Black Americans will not forget Trump's behavior that runs counter to all the positive policies he is proposing just weeks before the election. Because of the lack of trust and his well-documented behavior, Black Americans cannot trust that there is any interest in following through with these plans. I was born at night, but not last night. This is simply a ploy to peel off some Black votes from Biden in key swing states. The Platinum Plan is a sign of political desperation, not commitment.

What Trump has done to Black Americans might be forgivable, but not forgettable. These things he cannot erase:

In the 1960s and 1970s, Trump and his father discriminated against Black renters and homeowners in New York. He was successfully sued by the Justice Department for violating Fair Housing Laws.

Trump took out a full-page ad calling for the death penalty for five innocent Black and Latino youths, the so-called Central Park 5. He never apologized to these young men once it was determined they could not have committed the crime.

Trump after the murder of Heather Heyer in Charlottesville in 2017 opined that there were "very good people on both sides."

Trump has shown disrespect for countless Black leaders, from Elijah Cummings, to John Lewis, to Ilhan Omar, not to mention former President Barack Obama, or his recent remarks calling vice-presidential candidate Kamala Harris a "monster."

Then there's the almost total absence of Blacks in positions of power within his administration, and please do not tell me about Dr. Ben Carson. Is he even still in Washington?

Ironically, about the same time as the Platinum Plan was released, Trump also issued an executive order declaring that no federal agency or department could engage in diversity and inclusion training and that federal contractors that attempted to introduce such concepts could lose their contracts.

And for good measure, Trump's embrace of militias and other violent White supremacy groups encouraging them to go to the polls to intimidate Black voters in this coming election is proof that he does not have Black American interest at heart.

The sad thing is, had Trump initiated the Platinum Plan in February 2017 instead of in September 2020, he might indeed have won the hearts and votes of a significant number of Black Americans. This late conversion without a hint of repentance for his past behavior rings hollow. Only the most gullible Blacks will fall for this. Because his Platinum Plan is contingent on his reelection and sticking to his promise, Stevie Wonder will be proven right again, "you haven't done nothin."

4.3 It's Time to Change the Way Elect Presidents

November 13, 2020

Here we are more than a week after the election and while according a recent poll only 3 percent of Americans think that President Donald Trump won the election, we still do not have a definitive result because of frivolous lawsuits claiming voter fraud and a petulant president who refuses to concede. It would be one thing if this was a rare moment in our political history, but it is not. Every four years, we go through this Kabuki dance where we know going in that the results in the popular vote may differ significantly from the all-important Electoral College vote. It is time to change our system of how we select our president, still the most powerful position on the planet.

Traditionalists are always reluctant to change institutions that go back to our founding documents, regardless of the original intent of or the motivation for those documents. In the case of the Electoral College, political historians are ready to admit that this system had two main reasons for being.

Some Founding Fathers feared that allowing voters to decide who the president is would turn over this important decision to citizens too unprepared and too unsophisticated to make this decision consistent with the principles embedded in the Constitution. Some Founding Fathers felt a populist despot could influence an ignorant citizenry resulting in the election of someone who would act against the Constitution. Sound familiar?

The second reason for the Electoral College was tied to the fact that the Constitution allowed Southern states to count African slaves as three-fifths of a person for the purpose of determining the number of representatives in the lower house of Congress. While Blacks could not vote, they provided the number of representatives that protected America's "peculiar institution." It was because the South could count slaves to determine seats in the House and the number of seats in the House determines the composition and number of votes in the Electoral College that the South could protect slavery and wield power greater than the number of Americans legally allowed to vote at that time. This system worked for the slaveholding south, and in particular, Virginia. For the first 32 out of 36 years of the country, the president of the United States was a slave-owning Virginian (Washington, Jefferson, Madison, and Monroe each served eight years). It is ironic that non-eligible Blacks determined our first presidents—addition by subtraction.

Given this history and the problems, the Electoral College has caused in recent years with Democrats winning the popular vote in five out of the last six presidential elections, but only winning the White House three times, it is time to innovate.

There are several proposals being considered right now to eliminate the Electoral College completely or change it to better reflect the popular vote. I have a suggestion that would allow us to keep the Electoral College but would simultaneously provide citizens a more direct say in who becomes president.

Under current circumstances, the presidential election has come down to the outcomes in just a hand full of states; the so-called "battleground" states. It is troubling to me as a citizen from the non-battleground state of Connecticut that on the margin, my vote is less valuable than votes in one of those battleground states. As a result, the candidates spend most of their time wooing voters in the battleground states, ignoring citizens in the non-battleground states, including some of the most populated states in the country, California, New York, Illinois, and until this year, Texas. This lack of attention is due almost entirely to the Electoral College.

A solution to this would be to keep the Electoral College but make an adjustment to the final vote counts in each state to reflect the national popular vote total.

I did the calculations. This year, the national popular vote margin was 3.4 percent in favor of President-Elect Joe Biden, 50.8 percent to 47.4 percent. If a 3.4 percent adjustment was made to the final vote count in each state, it would have no impact on the allocation of Electoral College votes in the deep blue or the deep red states, but it would be decisive in states where the margins were small. Making this adjustment would have put Michigan, Nevada, Wisconsin, Arizona, and Georgia squarely in the Biden column. These are states that after the recounts, Biden will likely win anyway. However, this adjustment would make these races beyond doubt. Importantly, this adjustment would have flipped North Carolina for Biden and given him a total of 321 Electoral College votes, an Electoral College victory more in line with his 5 million-plus popular vote victory. We would have known who won the election by 11 p.m. Election Night!

This simple innovation of valuing the popular vote would provide all Americans an equal contribution in the outcome. This would also force candidates to pay attention to the entire electorate and not just a few battleground states. President Donald Trump made 14 trips to Pennsylvania while Biden was in the Keystone State 17 times. How many times do you think these candidates would have visited and fought for votes in the nation's largest states if the popular vote had an impact on the outcome?

It is time to innovate our presidential election process and let the results of the national popular vote count.

4.3.1 Critical Race Theory and the Teaching of American History—Why the Fear?

July 9, 2021

"***We think ... that [black people] are not included, and were not intended to be included, under the word "citizens" in the Constitution, and can therefore claim none of the rights and privileges which that instrument provides for and secures to citizens of the United States.***" Supreme Court Justice Roger Taney, Scott v Sanford, aka. The Dred Scott Decision 1857.

Late last month, the U.S. House of Representatives voted 285–120 to remove the bust of Supreme Court Justice Roger Taney whose words in the 1857—Dred Scott decision in effect made slavery legal everywhere in the United States. It took a Civil War and the 13th and 14th Amendments to the Constitution to correct what most legal historians agree was the worst Supreme Court decision in the history of the Court. And it took over 150 years after the Dred Scott decision for Congress to realize that honoring this Justice was an affront to many citizens, particularly the descendants of former slaves.

This summer, the culture warriors on the political right have focused their ire on the teaching of Critical Race Theory (CRT). CRT has been attributed to the work of legal scholars, including Derrick Bell, Richard Delgado, Patricia Williams, and others. The principles of CRT, however, go back to the writings of W.E.B. Dubois (1868–1963) and Frederick Douglass (1817–1895). CRT's primary point is that institutions, legal, economic, cultural, and political, have tremendous negative consequences on the lives of Black and Brown people.

I suspect the purpose this focus on CRT is Republican fear of losing democratic elections. These critics know that for them to have a chance of solidifying power they need to inspire their base to have a reason to vote in the 2022 mid-term elections when Donald Trump will not be on the ballot. They are hoping that by highlighting the dangers of CRT they can maximize their voter's participation.

But even if you take them at their word, that this is about content not politics, their criticism of CRT is baseless.

For every theory, there are competing theories. One of the competing theories of the anti-CRT crowd is that American institutions are (or should be) color blind; that life outcomes are determined by natural ability; that America is a meritocracy where the cream rises; that those at the bottom of society are there because they have not tried hard enough while those at the top are deserving of the rewards they have "earned" because of their superiority; and finally, that attacking American institutions is attacking the goose that lays the golden eggs. To critics, CRT is fundamentally anti-American.

The critics of CRT also appear to be almost exclusively White Americans. While it is unknown what percentage of White America is hostile to CRT, it follows that the critics of CRT also believe CRT to be anti-White. To these critics I say, it is possible to believe that American institutions were and unfortunately continue to be rigged against Black Americans while at the same time believe that Whites are not inherently evil and the sole cause of the problems impacting Black lives. Personal responsibility and institutions are both important determinants of life outcomes for Whites as well as Blacks. We need to address both the institutions and personal behavior.

Names of theories can be as powerful as the thoughts contained in those theories. I am not suggesting renaming CRT, but an alternative name for CRT is American History. American history is the story of what happened, who participated, and how our institutions laid the foundation for that history and in many cases, determined those outcomes.

Critics of CRT have organized across the country to keep CRT out of public schools and to punish and threaten teachers and administrators who teach CRT to impressionable young minds. This attack on CRT is understandable if the purpose of education is to indoctrinate not to educate. No serious educator is advocating for teaching only CRT, but that is the unreasonable fear critics of CRT seem to believe—in cult-like fashion.

I wonder if the critics of CRT would ban the teaching of the role the Dred Scott decision—that forced an enslaved man—Dred Scott—to be returned to his "owner" even after he had moved to a territory that banned slavery. What are we teaching our children when it comes to understanding the causes of the Civil War and the role of human slavery? And more importantly, how American slavery continues to impact our lives today.

Would CRT critics prohibit teachers from exploring the history of discrimination in housing through the practice of redlining? Would they ban teaching the history of police murders of Black citizens? Would they ban teaching the post-Reconstruction history of voter suppression and nullification? Would they ban the teaching of the history of Jim Crow laws?

Our responsibility as teachers is to tell historical truths, no matter how painful or how incongruent that history is compared to a mythology that we sometimes prefer. We are a great nation with great people. Yet our past is not perfect, nor is our future likely to be perfect. Only by teaching history in its fullest, can we have the informed citizenry we need to make this a more perfect union.

4.4 White Grievance

February 17, 2022.

After the Civil War, Union troops remained in the states of the former Confederacy to enforce the post-Civil War Amendments; the 13th Amendment which made slavery illegal in the entire United States; the 14th Amendment which gave citizen rights to the newly freed slaves; and the 15th Amendment that provided equal civil rights for newly freed slaves. The former Confederate states had to agree to these Amendments and pledge not to be insurrectionists against the United States in order to rejoin the union. Soon after the War, several former Confederate states changed their state constitutions to reflect the equality expressed in the newly passed Amendments. Mississippi, the heart of the Confederacy, passed one of the most progressive State constitutions in the country, granting all citizens of the state, Black and White civil and voting rights.

The period between 1868 and 1877 was a period when Black former slaves were elected to Congress, mayors' offices, and the courts because of the rights granted in the federal Amendments and the new progressive state constitutions. This was the period known as Reconstruction.

But this period did not last long. Almost immediately following the passing of these Amendments and progressive state constitutions across the South, former Confederates banded together and terrorized the newly freed slaves with lynchings and pogroms and the era of Jim Crow was born. Jim Crow lasted from about 1875 to 1965. It was a period characterized by forced racial segregation, discrimination, and terror. It was during this period that I was born in Arkansas, a former Confederate state.

In Mississippi, the White terrorists took control of levers of power and re-wrote the Reconstruction era constitution. The new Mississippi constitution instituted poll taxes that effectively disenfranchised all Black Mississippians. Additionally, the new constitution created a laundry list of minor crimes that took voters off the rolls, thereby beginning the period of mass incarceration. And the new constitution created the legal separation of Black children from White children in public schools.

This history of Jim Crow is important to understand today because it demonstrates the power of White grievance that we are witnessing today. Rolling back voting rights, gerrymandering districts to reduce representation, banning books, and claiming insurrection is legitimate political expression are how these grievances are being manifested.

Let me first state, I do not claim that these are the opinions or feelings of most White Americans, but they appear to be at a minimum, the views of a significant minority of White Americans. And I also understand that people from all races, ethnicities and parts of the world have their share of scoundrels, pathological liars, sociopaths, and just bad people, so this should not be interpreted as all White people are bad and all Black people are good. I know better and so does everyone else.

As a Black American, I try to understand the source of these grievances. Black households have one-tenth the wealth of White households. Black men are more than 5 times more likely to be incarcerated than White men. Blacks suffer from higher death rates from COVID, and other diseases compared to Whites. Blacks live about ten years less than Whites. Black children are much more likely to receive inferior education than White children. Blacks are underrepresented on corporate boards of directors. Blacks are underrepresented as judges, doctors, and elected officials. Black businesses are less likely to get financed by banks, venture capitalists, or government loan programs. All of these disparities are today, not 100 years ago.

Despite these challenges, Black Americans have made great progress since the late 1960s. Millions of Black Americans have entered the middle class. Blacks have achieved success in higher education, law, business, politics, sports, entertainment, and philanthropy.

The grievances of some White Americans seem to be that their view of White superiority conflicts with the successes of Black Americans. The fear that they will not maintain their economic and political superiority over Black Americans was made palpable by the election of the nation's first Black president Barack Obama. Had there not been a President Obama, there might never have been a President Trump.

Former President Trump represented for aggrieved White Americans a return to an America—Make America Great Again—that was based on a racial hierarchy that was like the Jim Crow South. In the Jim Crow South, there was no Black person, regardless of their intellect, character, and achievements who could be superior to the poorest, most ignorant, most morally decrepit White person. All White people were superior to all Black people. The breakdown of this destructive fantasy is the source of aggrieved White Americans today. In Charlottesville, the racists chanted, "Jews will not replace us." It is safe to infer that their biggest fear was that Blacks might replace them.

There is the battle line facing the two views of America's future. We will either have a multi-racial, meritocracy, or we will have an America where race determines a citizen's ability to achieve their human potential. There is no middle ground in these two visions of a future America. It will be either one or the other.

White citizens today who do not embrace the aggrieved vision of America must double down on building a multi-racial, multi-ethnic democracy that acknowledges all of us have the potential to be great while celebrating our differences. Black citizens also have a responsibility to strive for greatness in a multi-racial, multi-ethnic democracy, despite the actions of the aggrieved.

4.5 Historical Roots of January 6

October 27, 2021

"All persons born or naturalized in the United States and subject to the jurisdiction thereof, are citizens of the United States and the State wherein they reside. No State shall make or enforce any law which shall abridge the privileges or immunities of citizens of the United States, nor shall any state deprive any person of life, liberty, or property without due process of law, or deny to any person within

its jurisdiction the equal protection of the laws." 14th Amendment (Section 1).

"They attempted armed self-defense in Colfax. The result was that on Easter Sunday of 1873, when the sun went down that night, it went down on the corpses of two hundred and eighty negros." John G. Lewis, legislator State of Louisiana, testimony at the Congressional Ku Klux Klan hearings of 1871.

As Congress and the American people brace themselves for what could be a soul-searching introspective, and divisive public hearing on the events of January 6, 2021, it is important to remember that this was not the first time there had been an attempt to overthrow democratic government. Soon after the end of the Civil War and the passage of the 13th, 14th, and 15th Amendments, there was a period of Reconstruction where with the assistance of federal troops and the efforts of White northern Republicans, democracy became a temporary reality for Black Americans. Blacks were elected to both houses of Congress and mayors of major cities throughout the South. This is not some archeological artifact.

Unfortunately, this period was short-lived. By the late 1860s, the Ku Klux Klan and White Citizens Councils ushered in a period of racial terrorism that lasted for the next 90 years. In this post-Reconstruction era, Jim Crow laws were passed in all the former Confederate states. These laws were not simply about public accommodation but these laws were designed to relegate former slaves and their descendants to a life of servitude, dependence, poverty, and political impotency.

But before Jim Crow laws were enacted, the KKK and their supporters terrorized, murdered, and stole the property of the newly freed slaves and their supporters. The 1871 KKK Conspiracy Congressional hearings are a massive volume of nearly 7,000 pages documenting the terror of the KKK throughout the South following the passage of the post-Civil War Amendments I highly recommend anyone interested in hearing what happened during this time as told by firsthand witnesses and victims read this important historical document. The quote above was from one of the most notorious racial pogroms in American history in Colfax Louisiana.

The KKK Terrorism Act was not effective in eliminating the Klan; however, it and the 14th Amendment are important American documents that can provide guidance to Congress and the American people as to what should be done about January 6. The KKK Act speaks directly to the desecrators of the Capitol, and more importantly, the leaders including the former President and others who orchestrated the events of that day and the actions that are being taken across the country to restrict the voting rights of American citizens.

The 14th Amendment clearly states in Section 3:

> No person shall be a Senator or Representative in Congress, or elector of President and Vice President or hold any office civil or military under the United States or under any State, who having previously taken an oath as a member of Congress or as an officer of the United States as a member of any State legislature or as an executive or judicial officer of any State to support the constitution of the United States shall have engaged in insurrection or rebellion against the same, or given aid or comfort to the enemies thereof.

If the January 6 committee finds that the events of that day was an act of insurrection against the United States Constitution, all of those who participated, planned, inspired, funded, and otherwise supported this action could be banned from holding public office.

The case can be made that the attempt of the insurrectionists on January 6 to stop and reverse the duly appointed electors for President and Vice President by violently halting the constitutionally defined electoral process is the definition of insurrection. The 14th Amendment advises us on what should be done.

You might ask, what does the KKK Terrorism Act have to do with January 6? The answer to this question is that violent efforts of organizations dedicated to the principles of White supremacy like the Proud Boys, the Three Percenters, and the Oath Keepers were active in this effort to overthrow the duly elected government. One significant image of that horrible day was that of an insurrectionist carrying the Confederate flag through the halls of Congress, something that never happened during the Civil War. It is clear the insurrectionists and their supporters were actively trying to suppress and nullify Black votes in several key cities and states. The insurrectionists of January 6 are the ideological progeny of the KKK. And just as their predecessors attempted to hide their true intentions of White supremacy under the guise of states' rights, these modern insurrectionists have the gall to justify their acts of insurrection as an expression of election security. Hogwash.

In my dealings with elected officials of all parties, I have found that many of them keep a pocket-sized copy of the United States Constitution and the Amendments on their person. For the good of the country, I think we all should read those founding documents that begin with We the People.

4.6 The Supreme Court, Affirmative Action in Education and How I Feel

June 29, 2023

Today, June 29, 2023, the Supreme Court in a 6 to 3 decision, with the vote taking place along party lines, ruled to eliminate race as an independent variable in public and private college admissions. The six Republican justices appointed by Republican presidents voted for the plaintiffs to eliminate affirmative action as we know it, while the three justices appointed by Democratic presidents voted against the plaintiffs. This ruling puts a nail in the coffin of a legal movement that began with the founding of this country and the original constitution's view that Blacks were not deserving of citizenship—that Blacks were not human, or at best only three-fifths human.

Almost immediately after the landmark 1954 Brown v. Board of Education decision, cries of reverse discrimination could be heard from Whites who claimed harm from Brown inspired programs and future Supreme Court rulings that supported Black students, businesses, and communities. Legal redress of historical racism never was given a chance, and it turns out today never had a chance of success.

America's racial sins have never been adequately addressed. I am writing partly out of rage, but partly out of the knowledge of an intellectual history that took place largely out of sight of White Americans. Black Americans have debated the issue of our relationship to America since we were captured in Africa and brought here as slaves forced to work to make White men rich on plantations and factories that can only be described as hell on earth.

Two schools of Black thought quickly developed in the early eighteenth century. Slave revolts in North Carolina, Virginia, and Louisiana were Black expressions of resistance and the demand to return to a state of freedom from Whites and freedom from America. This school of thought was also encouraged by the only successful slave revolt in the history of the world that took place in Haiti in 1804, not thirty years after the successful American Revolution. Haitian revolutionary violence kicked out the French and directly led to America's first major expansion—the Louisiana Purchase. The defeated Napoleon needed the money to continue his European wars of expansion and had to sell land that today makes up 15 states, Louisiana, Missouri, Iowa, Oklahoma, Kansas, Nebraska, North Dakota, South Dakota, and parts of Minnesota, New Mexico, Montana, Wyoming, and Colorado. These red states and their citizens should thank Black Haitian revolutionaries for their lands and freedom today.

This revolutionary spirit continued through the work of Marcus Garvey, Stokely Carmichael, and Malcolm X. The move for independence from colonial powers in Africa culminating in the election of Nelson Mandela in South Africa in 1994 was directly tied to Black American and Haitian nationalism. There is historical documentation that American slaveholders, including the founding fathers were profoundly afraid of the impact the Haitian Revolution could have on American slaveholding states and the republic still in its infancy. In 1800, slaves in South Carolina comprised most of that state's population.

The nation fought a Civil War over slavery. But the question of what to do with the former slaves was unresolved then and unfortunately is unresolved now. The opportunity was to pay reparations. To provide land for the former slaves, to even give former slaves entire states.

Black freemen before the end of the Civil War and after the war debated the options of returning to Africa—some did, and Liberia was born. Some debated staying in America and becoming fully integrated into American society. It was on this issue that the two dominant streams of Black thought emerged at the beginning of the twentieth century. Those two schools of thought were led by Booker T. Washington and WEB Dubois, respectively. Washington expressed the view supported by northern industrialists that the two races should be separate on social and political issues but integrated on economic issues. He expressed his views at the 1895 Atlanta Exposition. WEB Dubois, the Harvard trained sociologists led a school of thought that wanted full integration of Blacks in all aspects of American society. Dubois was one of the founders of the NAACP. And it is historical irony that his alma mater was a defendant in today's ruling.

Since 1954, with a brief interlude when Stokely Carmichael, Malcolm X, and the Black Panther Party preached separation, the dominant view among Black Americans was that the only logical path was full integration into American society with help from the federal government, particularly the Supreme Court. With today's ruling all hope of that is dashed. The Supreme Court under the guise of race neutrality has sided with the racists. Arguing how could it be racists if Clarence Thomas supported the majority, after all he is a Black man. Clarence Thomas' skin color does not make him a Black man.

The question now is: What comes next? My crystal ball tells me that large numbers of Black Americans and allies in all races and ethnicities will attempt to fight this ruling by coming up with "workarounds" to mitigate its impact on college admissions. But I feel even larger numbers of Blacks born in this country will seek alternatives that look like the solutions proposed by Malcolm X, Marcus Garvey, and Abraham Lincoln. I am afraid this ruling has unleashed racial chaos without any solutions provided by the Republican justices.

4.7 The Descendants of Stolen People

July 28, 2023

Ron DeSantis could have let things lie in intellectual purgatory where Black history is celebrated without the pain of looking at the implications of that history on the living. Purgatory is a state that is understandably unsatisfying, yet better than one of the alternatives. The resulting reaction and firestorm led by President Joe Biden and Vice President Kamala Harris and every scholar of American history is best described as severe condemnation of the ignorance, malice and destructive power of Florida's actions and DeSantis' words. I do not think it is overreacting by saying this fight to protect the truth is a fight a democratic society cannot afford to lose and must be fought until the racists submit to unconditional surrender. The war of ideas should not be and need not be violent, but make no mistake, canceling Black history is one of the most violent acts that can be committed. Nazi Germany, Pol Pot Cambodia, and Rwanda demonstrate that the erasure of history is dehumanization, and this is often followed by physical violence and genocide.

This battle is taking place in the context of other battles that threaten Black economic progress and bring into question what the best strategy for Black Americans is. We need allies and we must be open to collaboration with others who seek justice and truth. Justice and truth should be what binds us in the pursuit of a better America. And we cannot forget that justice and truth are color blind.

The reassessment of being Black in America starts with the question of who we are and how we want to be addressed. We have gone from being called the N word and other derogatory names designed to make us seem less than human to Negro, Black, African, Afro American, African American, Minority, People of Color, and BIPOC. Unfortunately, none of these seem to accurately describe our relationship to America or are universally accepted by the people being described.

I want to suggest an alternative name that describes and honors our past—the "Descendants of Stolen People." Every person, except for recent immigrants from Africa, who can trace their ancestry to Africa are descendants of stolen Africans. This is a far superior term than any of the current terms used to describe or categorize us, particularly the term "minority."

The term minority is more a political description than a racial/ethnic one. It is also a term that will eventually be flipped on its head. The Census Bureau predicts that by 2060, most Americans will not be of European origin. Will that make Europeans the new minority? Think of all the organizations, like the one I led, the Greater New England Minority Development Council, whose name will be confusing at best as we approach 2060.

We know from the experience of South Africa that minority rule is possible and is usually accompanied with violence against the majority to maintain minority political power. America is facing the reality of White minority rule in our lifetimes. The Justice Department recently ruled that Alabama's redistricting needs to be redrawn to create a majority Black district. The Alabama legislature and Governor are fighting this order. Voter suppression is rampant across America, and the intent is clearly to maintain political control by the current majority who will in coming decades become the minority.

The term "minority" has no probative value when it comes to race and ethnicity and therefore to policies, programs, organizations, and government. We need a new language. Words have power and I recommend that we stop using the term "minority" to describe the Descendants of Stolen People. The Descendants of Stolen People is a unique identifier and brings into clear focus how we became "Americans." So please, do not call me minority, or Black, or African American. I prefer to be called, a Descendant of Stolen People or DSP for short.

In Canada, every public meeting begins with what is called the Land Acknowledgement. This week I presented to a group of Canadian supplier diversity professionals and before we began, the organizer of the meeting read the following statement:

> I'd like to begin by acknowledging the Indigenous Peoples of all the lands that we are on today. While we meet today on a virtual platform, I would like to take a moment to acknowledge the importance of the lands, which we each call home. We do this to reaffirm our commitment and responsibility in improving relationships between nations and to improving our own understanding of local Indigenous peoples and their cultures. From coast to coast to coast, we acknowledge the ancestral and unceded territory of all the Inuit, Métis, and First Nations people that call this land home. Please join me in a moment of reflection to acknowledge the harms and mistakes of the past and to consider how we are and can each, in our own way, try to move forward in a spirit of reconciliation and collaboration.

There is an active and serious discussion in Congress and in several states about the possibility of reparations to the Descendants of Stolen People. In addition to the compensation owed DSPs, there also needs to be an official acknowledgment of the harm done to DSPs, like the Canadian acknowledgment that Canada was occupied by others before the Europeans arrived.

We need an acknowledgment of the harm perpetrated against DSP. This acknowledgment is a necessary start to reconciliation between the descendants of the American Founders and the Descendants of Stolen People. Because even if wealth could be or will be transferred to DSPs in the form of reparations, without a public acknowledgment, the healing of our nation cannot take place.

The implications of how we define ourselves are profound. We should start today. Think about how "Black History" would be taught in DeSantis' Florida if it were called "the History of the Descendants of Stolen People." That is who we are.

4.8 Is America a Racist Country?

January 30, 2024

E Pluribus Unum—Out of Many One

Recently, CNN host Jake Tapper asked Nikki Haley, a candidate seeking to represent the Republican Party in the 2024 presidential election, "Is America a racist country?" This is not a tricky question. It is not a gotcha question unless you make it one. Her answer became news because she flunked this oral exam. Gotcha!

Her response and her body language demonstrated to any voter who understands the seriousness of the job of president of the United States, that this is not a job for her. I cannot remember another Republican presidential primary where there was nothing to choose from other than intellectual mediocrity.

Pardon me for my pretentiousness, but a qualified Republican candidate for the leader of the free world would know the answer is embedded in a non-superficial understanding of actual American history—not the sanitized version Haley apparently and unfortunately received in the "segregation school" she attended in South Carolina, or the "anti-woke" version of history DeSantis wants to institutionalize in Florida's public schools and universities.

American history is filled with the distinction between American ideals and American practices. The lofty and revolutionary concept that all "men" are created equal made an explicit distinction between the genders. And make no mistake, it was not an oversight. For the first 144 years, after the signing of the Declaration of Independence, women could not vote until the passage of the 19th Amendment in 1920.

The dissonance between principles and practices was even more acute when looking at the plight of Black Americans. The original Constitution never uses the words slave, but says the following in Article 1, Section 2:

> Representatives and direct Taxes shall be apportioned among the several States which may be included within this Union, according to their respective Numbers, which shall be determined by adding to the whole Number of free Persons, including those bound to Service for a Term of Years, and excluding Indians not taxed, three fifths of all other Persons.

The so-called "three-fifths" clause effectively gave increased Congressional power to the southern states where slaves made up a sizable portion of the population, and which had smaller White populations than several of the northern states. Demographers estimate that 54 percent of the South Carolina population was Black in 1780.

Notably, women were included in the determination of how many representatives a state could send to Congress, even though they could not vote for those representatives. The three-fifths clause was not edited to include Blacks as full citizens until the post-Civil War 13th Amendment.

The Supreme Court's 1954 *Brown v. Board of Education* decided racial discrimination in public education was unconstitutional and that public schools should be integrated with "all deliberate speed." Soon after *Brown*, private segregated White schools sprung up in the South like mushrooms on a cool summer morning. Southern cities and towns decided that instead of allowing Black students in their public schools, they would create their own private schools, letting in who they wanted on one hand, and excluding who they did not want on the other. Nikki Haley attended one of these "segregation schools."

Since White kids would be taken care of in private segregated schools and White people controlled local government, public education became underfunded, in effect taking money from the public schools to help White people finance their private segregated schools.

America keeps providing us stories like Haley's that for Black Americans reminds us how strongly racism continues to play a role in our lives and in American society.

Nikki Haley is the daughter of Indian immigrants who practiced the Sikh religion. Her father Dr. Ajit Singh Randhawa was a biology professor at Voorhees College, a historically Black college in Denmark, South Carolina. It strains credulity that she never witnessed (or experienced) racism in America.

Chances are a turban- wearing professor would have had a difficult, if not impossible, challenge finding a faculty position at the University of South Carolina (or any other White university in the Southern states). It was not until 1963 that the University of South Carolina admitted its first three Black students. This was not random. After the Civil War, the insurrectionists seized power and created two separate and unequal systems one White and superior, the other Black and inferior, each covering every aspect of life including where you were born, which school you attended, where you could live, where you could sit in public transportation, where you could eat, where you could drink from a water fountain, where you could go to the bathroom, and even where you could be buried.

I find it hard to believe that Nikki Haley did not see the remnants of Jim Crow when she was growing up and attending schools set up by slaveholders for their privileged White children. I also find it hard to believe that an intelligent man like Dr. Randhawa would not do what Black parents have done for generations—tell their kids to be the best they can be and admonish them that they must be twice as good as the White kid to be treated as an equal. When this narrative works, most successful Black Americans know their success is rooted in the struggle of past generations. Any Black person who denies the institutional barriers that exist are frequently viewed as misguided or a traitor to the race. The tell-tale sign of this betrayal is the claim that their success is primarily the result of what they did and nobody else.

Haley had a golden opportunity to tell the real story of who Nikki Haley is by answering Tapper's question. Could she have enhanced her credibility if she had answered the question with a story of her mother and father, or stories they told her about Sikhs in India or being Sikh in America, or stories about the Black students her father taught at Voorhees College? She could have reached countless Black voters, or at least gained their respect.

Haley knows that America has a racist history and unfortunately there are legacies of those racist systems that influence the lives of millions of people today. There are scores of stories she could have told Tapper. Whether she was trying to play it safe, pandering to history revisionists, or simply lying, her answer was an unforgettable *faux pas*.

4.9 Run Kanye Run

September 17, 2020

My first reaction to Kanye West's quixotic run for the presidency brought back memories of Jill Stein's similar run for the same office in 2016 as the candidate of the Green Party. Stein did not have a snowballs chance in you know where to win. Her stated goal was not to win, but to bring attention to the Green Party agenda, but the real goal was to provide disgruntled Bernie Sanders voters a place to go when he lost the nomination to Hillary Clinton to elect President Trump. Stein selected a black running mate with an Afro-Centric name Ajamu Baraka as her vice-presidential running mate.

This time around we have a well-known successful Black rapper with a history of bi-polar disorders, running as an independent with a White woman Christian fundamentalist as his running mate, Michelle Tidball. I bet even political junkies will not remember her name after the election.

In 2016, Jill Stein was able to win 51,000 in Michigan. Donald Trump won Michigan by 11,000 votes. In Wisconsin, Jill Stein won 31,000 while Donald Trump's margin of victory was 22,000. Now we don't know where those votes would have gone if Stein had not been in the race, but it is not just speculation that her role in the race had the same effect in 2016 that Ralph Nader had in 2000 when Nader won 97,000 votes in Florida and George W. Bush won the Supreme Court contested Florida election by a mere 537 votes.

So, third-party candidates who have no chance of winning can have a disproportionate impact on the election results in tight races. It certainly looks like 2020 could be a tight race.

What also makes Kanye's run so suspicious is like Stein who had Russian connections; she visited Russia in December 2015 and sat at the same dinner table with Russian president Vladimir Putin and future, but short-term National Security advisor Michael Flynn; Kanye has strong connections to Trump. We will have to wait to see if Vlad starts sporting a Ye t-shirt.

Kanye West's entry into the race also raises interesting suspicions because Trump lawyers were the same lawyers used by Kanye West to try to get on the 2020 presidential ballot in Wisconsin. Smells fishy to me.

I was having a conversation about this recently with my friend, Professor Gerald Jaynes of the Yale Economics Department. His take on this was very interesting and convincing. Unlike Jill Stein, Kanye West's run for the Oval Office is likely to attract some Black conservative voters who are fans of Kanye's Christian rap, but maybe not so much as his earlier collaborations with Jay Z. The intersection of black conservative Christian rap fans has got to be an infinitesimally small portion of the electorate. And if anything, this group of marginal Black voters is likely to lean more heavily towards being in Donald Trump's camp and not the Joe Biden camp.

Therefore, supporters of Biden should not think history is repeating itself. If anything, Trump's clandestine support of Kanye is good for

Biden. These are voters who would never vote for Biden anyway. So, their vote for Kanye is one less vote for Trump.

Somehow, I do not think Trump has figured this out. I think he is so unidimensional in his thinking that he thinks Kanye equal Black, and therefore, he will steal black votes from Biden. Trump probably thinks Kanye can be his 2020 Jill Stein.

Remember that incredibly strange meeting Trump had with Kanye in the Oval Office. I suspect that this is when Trump came up with this idiotic strategy. According to Michael Cohen, Trump has said to anyone who will listen, that Blacks and Hispanics are too stupid to vote for him. He must think the best thing to do under these circumstances is to give these voters someone they can support who will help him get reelected.

I have news for President Trump; Black voters are not stupid. We know that Kanye has no chance of winning. We know that this is a fraudulent attempt to split the Black vote. We know that Kanye is even less qualified for the presidency than you, the most unqualified President in history, and look at what this has wrought. Black voters know all too well from our history that when it is convenient for Whites in power, they will use other Blacks to suppress the legitimate goals of the Black community. There is a term for these tools—Uncle Toms. President Trump this won't work. In fact, it will have the opposite effect.

All I can say is run Kanye run.

4.10 The Meaning of "Stand Back and Stand By"

October 3, 2020

In the first 2020 presidential debate that looked more like a car wreck on I-95 complete with fire, smoke and emergency vehicles with flashing lights, candidate Trump not only refused to condemn the violent and racist para-military group, the Proud Boys, he encouraged them to "stand back and stand by." President Trump views this violent fringe group, with roots in the same sociopathic thinking of Oklahoma City bomber Timothy McVeigh as pro American and supporters of law and order. That act of domestic terrorism on April 19, 1995, killed 168 fellow Americans and wounded 680, including innocent toddlers enrolled in a day care center in the building. Is it any surprise that Americans of goodwill are being called to reject the provocations of the sitting president of the United States who is encouraging voter intimidation by this vigilante proto-terrorist organization? This is not just my opinion, but it is also the view of FBI director Christopher Wray.

The question President Trump's comment raises is: What are they afraid of? It is clear from all of what he says and all of his administration's actions, including building a wall between Mexico and the United States, that the overriding fear animating this abhorrent behavior is the fear of becoming a racial minority in a country that he and some of his supporters seem to think should be dominated economically, politically, and demographically by Whites, forever.

In February, the Census Bureau reported that the number of non-Hispanic Whites will decline from 198 million people to 179 million in 2060. Over the same period, the Black population will grow from 43 to 60 million. Hispanics in America will grow from 57 million today to 111 million in 2060. The Asian population will double from 18 million today to 36 million in 2060. The White non-Hispanic share of the U.S. population will decline from 61 percent today to 44 percent by 2060. It is a common refrain that demography is destiny. But even by 2060, White non-Hispanic Americans will continue to represent the largest single demographic group.

President Trump and his White supremacist supporters fear these inevitable demographic shifts in the composition of American society. White privilege is something with clear economic and political value for Whites but has costs for everyone who is not White. A superior privilege that every American should experience is the privilege associated with being a citizen. While we know we have not lived up to the ideals in our founding documents, "that all men are created equal," the language of those documents was changed so that today there are no qualifiers that proclaim, all White men are created equal and all others (including White women) are inferior.

Our society is evolving. Change is hard, particularly for those who are ill-equipped to adapt to new circumstances. The Proud Boys and other violent extremists, contrary to basing their actions on feelings of superiority, are demonstrating by those actions and words a profound sense of inferiority rooted in questions about their ability to adapt. Their claims of racial superiority are ironically the result of feelings of inferiority. Guns and pseudo-military clothing and flags and other symbols of White racial solidarity will not be enough to turn back the tide of the coming demographic shifts or peacefully address these feelings of inferiority.

White Americans, and I emphasize the word Americans, should embrace these demographic changes, and not as threats to their status or sense of economic security. The political motto of the United States, E Pluribus Unum, Out of Many One, was suggested originally on July 4, 1776, and placed on the seal of the United States of America. That motto has more relevance today than ever, and it should be our guide.

We do not need this president to tell the Proud Boys to stand by. In fact, we do not need the Proud Boys or other racist groups to educate us on the principles of democracy and citizenship. No one is questioning their rights to bear arms, although that right is literally second to the rights of political freedom. The most important of those political rights is the right to vote without fear or intimidation.

The president is playing a dangerous game telling the Proud Boys to stand by. He has proven himself incapable of rational and principled actions, but we must not allow him to destroy, or encourage the destruction of, American democracy. We must vote him out. We deserve better. America deserves better.

4.11 The Threat of White Minority Rule in the United States

October 14, 2021

Cheating in politics, like baseball, is a national pastime. Major League Baseball is wrapping up its fall classic with an American League championship between the cheating Houston Astros versus the Boston Red Sox, whose manager was found guilty of cheating while a coach with the 2017/18 cheating Astros. In politics, the Republican Party is trying to cheat its way into power and create White minority rule. Just as baseball had to rewrite and enforce its rules to keep up with the creativity of rule abusers to protect our political processes, we must update our political rules. If we don't make changes, we will end up with White minority rule, like what took place in disgraced apartheid South Africa.

Updating the rules is something that America needs to consider when it comes to our election processes, but not in the way most Republicans have been advocating. Republicans are trying to change the election rules to prevent "widespread voter fraud" that even the highest-level federal election officials in the Trump administration stipulate did not happen.

Nor did those stealthy Cyber Ninjas, despite months of looking in Maricopa County Arizona, find anything other than a larger margin for the winning candidate, Joe Biden.

Democrats need to rewrite the rules to prevent Republicans from actively stealing the 2022 and 2024 elections. But it is not just Democrats who need to raise the alarm, all Americans interested in Democracy need to clearly be on the side of free, fair, transparent, and open elections.

According to the National Conference of State Legislators, Republicans currently control 54% of all state legislative elected offices with Democrats controlling 45%. In twenty-three states, Republicans control both houses and the governorship. In 16 states, Democrats control both houses and the governorship, and eleven states there is divided government.

If 2020, the 16 solid blue states with 195 electoral votes all voted for President Biden. In 2020, the 23 solid red states with 221 electoral votes all voted for Trump except, Georgia and Arizona (surprise-surprise). Twelve states with split control of the legislature and the governor's office had 122 electoral votes. Biden won 81 electoral votes from those states and 41 went for Trump. Included in those politically divided states are Michigan, Wisconsin, and Pennsylvania with a total of 46 Electoral College votes—all three of those states went to Biden.

The game plan for the Republicans is as clear as day, restrict voting in Michigan, Wisconsin, Pennsylvania, Arizona, and Georgia and if they are successful limiting or nullifying votes in those five states and they hold the states they won in 2020, it would give the Republicans the White House in 2024.

Plan B for Republicans is to change the laws so that the Secretaries of State in those five states can overturn the results of the election by claiming voting fraud. Voter suppression and election nullification are the cheating game that is taking place right in front of our eyes.

An even closer look at the Republican strategy reveals a very clear racial strategy to suppress Black and Brown voters. The five key states are also home to cities with large populations of Black and Brown voters—Atlanta, Philadelphia, Detroit, Milwaukee, and Phoenix. The average non-White and Hispanic population in those five cities in 2019 was 65.2%. It was in those cities/counties where Trump lawyers were thankfully thrown out of court. For Republicans, it is not about voter fraud. It is about suppressing Black and Brown voting in these critical cities and winning at any cost.

What is the solution?

Some are calling for independent election monitors and arbiters—kind of like what the UN does in countries without long democratic traditions. Umpires are absolutely needed in baseball, and I suspect they are needed to keep our elections clean. However, I do not believe Americans would ever stand for a committee of Americans not to mention UN appointed observers supervising our elections.

The more realistic solution to counter Republican efforts to steal future elections is to get massive voter registration and turnout, particularly in these five urban centers. In 2020, non-Hispanic Whites voted at 71%, Blacks voted at 63%, Asian Americans voted at 60%, and Latinx voters voted at 54%. Biden won the 2020 election by over 7 million votes. If the rate of participation by Black, Asian and Latinx voters increased to just 65%, the national vote would make it all but impossible to steal the election. This level of participation would give Democrats a Gavin Newsom size win in 2022 and 2024 and eliminate any doubt about the outcome.

The key to fair elections is participation. Without participation, I am afraid we are looking at an American future of White minority rule. Minority rule is inconsistent with democracy and minority rule based on race is a recipe for national disaster. If baseball can enforce fair games, we should be able to enforce fair elections. Former President Trump repeatedly claimed the election was rigged. The way to fix that is for Americans to vote in record numbers.

There are two diametrically opposed potential futures. One is White minority rule. The second is multi-racial, multi-ethnic democracy. I hope we choose the latter.

4.12 The Market for Lies

March 16, 2023

"A lie can travel around the world and back again while the truth is lacing up its boots." Mark Twain.

As an aging economist whose tools and abilities are not what they once were, the media crisis of the century, (the Fox News, House Republican Leader McCarthy and Donald Trump crisis) lends itself to an economic analysis of lies. There is unquestionably a market for fiction that sordidly cloaks itself as fact, that is, "alternative facts" Analyzing any market, you start with the fundamentals of demand and supply. These simple tools can inform us on what is happening and the likely outcomes in this market for lies.

The demand for lies, particularly political lies, must be understood by looking at what economists call utility functions. Economists take it for granted that every individual seeks the maximum amount of happiness, or what economists call utility. The problem is our resources are insufficient and our desires for more stuff are unlimited. Economists also assume everyone is different in the things they desire. My happiness might come from being on a beach in Montego Bay. Your happiness might be skiing down the Alps, or simply feeding your family. With these simple assumptions that underpin human behavior, the question is: Why would anyone demand to be lied to?

When you lose and you think you won and there is someone else out there supporting your false claim of victory, you will demand, not just more, but ever-increasing amounts of this hogwash. Believing in lies requires repetition and increasing levels of message intensity. This is how Fox News boosts the demand for lies, literally growing the market for lies.

When the millions of middle- and lower-income White Americans latch their wagon to a snake oil salesman because he convinces them that their problems are not with economic factors that have created their environment, but with Blacks, Hispanics, immigrants, gays, Jews, anyone different looking or not believing like them. The demand for being lied to is the result of deep and real harm caused by global economic forces. It is this harm that feeds the racism, anti-LGBTQism, and anti-Semitism. The current governor of Florida is promoting his own new set of lies related to wokeness, but this too will fail, crushed under the weight of untruth. Desantis's hogwash is the result of the same factors that led to belief in the "big lie."

The supply of any good or service in capitalist markets is driven by the desire for profits and wealth. Individual consumers sometimes tell the truth and sometimes don't. In this way, all of us are potential suppliers of lies. And now with social media, we can broadcast those lies as fast as Twain predicted. But the shocking nature of this recent week's event around revelations disclosed in the Dominion Voting Machines defamation case against Fox News is as earth shaking in the media world as January 6 is in the political world.

There are big generational events like Pearl Harbor, the assassinations of JFK, MLK, and RFK in the 60's, 9/11, and more recently, January 6 that defines the times. The Dominion defamation case might be just as big as these seminal events, but it is fundamentally different because in Dominion, the news is the "news." Fox's deceit, hypocrisy, and maliciousness are direct counters to how the press is supposed to work, particularly in its relationship with political news. "We report—you decide" became we decide, and we fabricate. Fox became an immensely profitable wealth generating machine in the business of producing and selling lies.

Fox owners and on-air personalities knew what they were saying on-air was lies. They broadcast the lies that Trump and his team also knew were lies, to millions of Trump supporters whose eyeballs are valuable to hundreds of American corporations, who also did not care what the lies were, so long as Fox viewers witnessed their commercials. In these symbiotic relationships between Fox, Trump, the Republican Party, and Fox's advertisers, there is no need to be overtly conspiratorial. This is a business model that works for all the players except the Fox viewers if the bubble does not burst. But all bubbles burst.

In 1634, in the Netherlands, tulip bulbs started going up in price so fast that by 1637, just before the bubble burst, the choicest bulbs would sell for what is equivalent to $1 million today. As is the case in any pyramid scheme, or "once in a lifetime" opportunity these events are driven by lies and they are all eventually exposed.

More recently, the Silicon Valley Bank collapsed in seemingly a nanosecond. Collapses with tulip bulbs or runs on banks happen because the truth is revealed and the hoodwinked try to salvage their dignity and or their assets.

Fox's market for lies is looking more and more like the insanity of the Dutch tulip bulb market than normal markets. Understanding this, I see a bursting bubble where Fox will lose large chunks of its audience because of its brazen hypocrisy and hubris. It is interesting that the tulip bubble lasted about as long as the big lie. Some forlorn Fox viewers will find new media homes supporting their delusions. Twitter, One America and News Max may attempt to fill the void, but millions will wipe the scales from their eyes and rejoin us here on Earth One and leave Earth Two and its alternative facts forever. We will always have a demand for lies, but if we are to survive as a democratic society, the cost of supplying and broadcasting those lies must be grave. History shows there is no avoiding the inevitable crash when behaviors are based on lies.

4.13 Blacks and the Republican Party

February 27, 2024

I do not know what political party my great grandfathers either belonged to or who they voted for, but it is likely that their sentiments, having been born in the middle of the nineteenth century, were with the party of Lincoln. This is particularly the case since they were either born before the conclusion of the Civil War or shortly afterward.

Former President Donald Trump recently addressed the Black Conservative Federation. His statements as to why Black Americans today should or will vote for him demonstrate his profound ignorance, hubris, and quite frankly, racism. The Republican Party as led by Donald Trump is not going to attract many, if any, Black Americans because he was legitimately impeached for attempting to steal the presidential election of 2020. Black Americans have a deep appreciation for the law. This appreciation did not and does not originate from American constitutionalists like Jefferson, Adams, Monroe, or others. Our deep-seated sense of justice, fairness, and law comes from our historical and unseen roots in the African traditions of Ma'at.

Black history month is ending, but before it does, all Americans should know more about Black history than just the history of slavery, the Civil Rights and Voting Rights movements, and the significant cultural contributions of the race. We need to go back much further in time to realize that Africans were the founders of civilization.

For some, this claim may be hard to digest. For others, this claim may fall on deaf ears. And for too many Black Americans and White Americans, it may violate everything they thought they knew about their culture and civilization. The foundation underpinning prevailing beliefs in European or White superiority is the misguided assumption that Blacks contributed nothing to the civilization of the planet. The fact that the commemoration of Black contributions is confined to one month a year is an indication of feelings of White superiority and Black inferiority.

I encourage my readers to study the ancient Kemet civilization. Kemet later became known as Upper Egypt. The first Pharaoh, Menes, also known as Narmer, was an African king of Kemet prior to his reign as Pharoah. Menes ushered in the first six dynasties of Egyptian Pharaohs, all of whom were Black Africans from Kemet. It was these Kemet Pharaohs who built the great pyramids of Giza—1,500 years before Moses; 2,500 years before Socrates, the Roman Empire and Lao Tsu;

3,000 years before Jesus Christ; 3,600 years before Mohammad, the Prophet; and 4,700 years before the formation of the United States of America. And it is part of the undisputed archeological record that all world religions are based on Kemetian religion and philosophy. Even the ancient Greek historian Herodotus knew this. Even the story of Christ is the retelling of earlier African stories of death and rebirth.

Black Egyptians gave the world advanced mathematics, medical science, architecture, writing, philosophy, engineering, and social justice. Ma'at was the Kemetian/Egyptian goddess in whose name the 42 principles of individual and social behavior were codified and were practiced throughout Africa, continuing even to this day. Ma'at represents balance, justice, order, and harmony. Trump represents chaos—the opposite of Ma'at.

The Black Republicans, to whom Trump addressed and made ridiculous and racist remarks, need to realize that today's Republican Party is not the Party of Lincoln. If Trump remains the leader of this party, and the party acquiesces in his toxic brand of leadership, those who continue to support him, Black, White, or any other ethnicity, should realize they are supporting an immoral, racist, and ignorant organization that will do significant harm to the interest of not just Black people, but all people.

If my great grandfathers voted Republican after the Civil War, it is understandable. If Blacks today vote Republican, they only demonstrate their ignorance of their own history and contributions to world civilization and their self-hatred. Blacks were the founders of civilization—not the criminals Donald Trump thinks we are. That is Black history!

4.14 The Coalition of the Avaricious, the Intolerant, and the Crazy

June 15, 2024

There is a parable about a poisonous snake that is trapped on a log in the middle of a flooded raging river when a beaver comes along easily navigating the turbulent waters. The snake, in the desperate fear of drowning, asks the beaver if he would ferry him to the safety of the banks. The beaver asks the snake: "Why should I trust you?" The snake promises the beaver he will not bite him. The beaver, with some hesitation, agrees to take the snake to safety. When the snake and beaver reach the river's edge, the snake bites the beaver. The beaver's last words before dying were: "Why would you bite me after I saved your life?" The snake's response was: "Because I am a snake; it is my nature."

When looking at the state of politics, this parable comes to mind as I try to understand the strange coalition of Donald Trump, the billionaires who financially support him, and the masses of disenchanted White American working-class citizens whose support make it possible for Trump to become the next president of the United States. In 2016 and 2020, Donald Trump asked the American people to elect him to the highest office in the world. Now, in 2024, he asks again, even though he has shown us his true nature over this nine-year period. What makes anyone think he will not bite America again in 2024?

The billionaires supporting Trump are doing so because they believe that their wealth—and the wealth of their families for generations to come is safer in the hands of a Trump presidency than in a Biden presidency, regardless of the impact of his policies on most Americans. They are happy to fund his super political action committees. To write a $1 million check for a billionaire is identical to a worker with $80,000 in annual income writing a check for $80. The billionaires' million dollars buys significantly more influence than $80 will ever buy for the Trump supporters who attend his rallies. The billionaires will get ambassadorships for their friends and family members, secretary and assistant secretary positions in government, and regulations that will make them even more money. Admittedly, if I could get those things for $80, I might sign up too. The avaricious have practical, albeit shortsighted, reasons for supporting Trump.

It has become a staple for leftists to say that White working-class Americans who align themselves with the affluent, business-friendly Republican Party are acting against their working-class interests. The counter argument to this thesis is that Trump supporters, like Nixon supporters before, see the liberal Ivy League elites as too far culturally removed from their day-to-day lives to work in their interests. But a deeper dive into this cultural argument reveals the underlying issue of race. Nixon's "southern strategy" was to cleave the White working-class from their traditional support for the Democratic Party by appealing to their racial interests. This positioning of the Democratic Party as the party for Diversity, Equity, and Inclusion and against discrimination was craftly rebranded by Republicans. Using tropes like the "Black welfare queen," despite the fact there are larger numbers of Whites receiving welfare than Blacks, and George H.W. Bush's portrayal of the threat of Black criminality to the safety of White women in the Willie Horton commercial, exemplify the point. The

Republican Party spoke to (and fostered) visceral, deep-seated fears of being replaced by criminals, lazy welfare cheats, and affirmative action beneficiaries like Barack Obama—a Democrat.

The crazy component of the Trump coalition would be the most entertaining if it were not so threatening. Some White working-class Trump supporters cloak their support in the raiment of a perverse form of Christianity. My understanding of this force multiplier is that God choose Trump to deliver America and the world from the evils of diversity, homosexuality, atheism, abortion, women's rights, and immorality. These "god-fearing" Trump supporters view Democrats as representing everything they are against. The irony, of course, is that their leader's life is the antithesis of the teachings of Christianity. In the strange logic of these supporters, God works in mysterious ways. It is interesting how forgiving they can be of Trump, when you do not see this level of forgiveness, or even tolerance, for humanity in general. This group of true believers is lost to reason.

Unfortunately, this is a powerful coalition of money and voters who will be with us regardless of who wins the 2024 presidential election. If the results are close, the billionaires, the racists, and the true Trump believers may never accept defeat or more appropriately, accept the will of the majority. This is a scary thought, particularly for racial minorities who, from the beginning of our time in America, had to abide by majority rule. If the Trump coalition cannot accept majority rule as reflected in a free and fair election, then that is an indication that their commitment to American democracy is limited by self-interest. They are demonstrating their nature and are willing to fatally bite and kill the American experiment, even if they get their way. Imagine what happens when and if they don't.

CHAPTER 5

Health and Education

Abstract Health and Education policy are key to the development of all Americans, but particularly Black Americans because of the legacy of slavery, Jim Crow, and discrimination. This chapter looks at how that history still impacts policy and outcomes today for Black Americans.

Keywords Health · Education · Brown v. Board of Education · Discrimination · Segregation · Health Disparities

January 17, 2020

5.1 Beyond Sheff: Innovative Solutions to Educational Performance in Segregated Urban Schools

This past week in Hartford, Connecticut, the Hartford Superior Court announced a settlement in the 30-year-old Sheff v. O'Neill school desegregation case. The settlement adds over 1,000 new seats to several Capitol region magnet schools and pledges to spend an additional $2 million to support this effort.

As a reminder, it was 65 years ago that Brown v. Board of Education was decided. The lasting sentence that captures the spirit of Brown

F. McKinney, *Critical Issues Facing America*,
https://doi.org/10.1007/978-3-031-93217-5_5

by Chief Justice Earl Warren (an Eisenhower appointee—oh how the Republican Party has changed) was: "Separate educational facilities are inherently unequal." Martin Luther King poetically stated, "Segregation is wrong because it is a system of adultery perpetuated by an illicit intercourse between injustice and immorality."

So, while this "settlement" may end Brown's legacy in Hartford, it does not and cannot address the fundamental changes that are necessary for the State of Connecticut and its cities and towns to address educational performance in still largely segregated urban schools.

Segregation in public education in Connecticut is a life-altering problem, and this settlement represents only a chink in the armor of segregated Connecticut. However, this decision will have only a marginal impact on segregation in the greater Hartford area, and more importantly, it will have no impact on reversing the fundamental problem facing African American children in urban communities, not just in Connecticut, but across the country.

Unsolved social and economic problems in African American communities have a way of finding their way into low-income White communities. The current rise in deaths of despair from suicides, alcohol, drugs once relegated to low-income black communities are now endemic in low-income White communities. Low-income Whites realize social and economic racial entitlements have limits. Low-income White Connecticut schools will begin to resemble low-income black schools. If you don't believe this, look at the similarities between the crack cocaine crisis that devastated black communities in the 80's and 90's and the opioid/fentanyl/meth crisis in White communities. Solving educational problems of performance in black communities can inform solutions to the social disintegration taking place in low-income White communities across the state and the nation.

Public education has failed African American children. It is not because teachers, administrators, and parents aren't trying. It is not because we have not spent enormous amounts of tax-payer dollars. It is not because we do not know that the system has failed African American children. The problem is the system is ripe for real innovation that addresses the unsolvable problems of educational performance within the current model of urban public education.

Public (and private) schools have been organized around a concept that children must fit the system, and not that the system should fit the child. Children start in kindergarten then progress through first grade, second grade, etc. After a year in each grade, at the elementary level, the child meets a new teacher. This system of handing off children each year to a new teacher is a counter-productive way to educate kids of all races, but particularly black kids from neighborhoods and families that are on the front lines of poverty. Psychologists agree that poverty traumatizes its victims, particularly the young. Teachers cannot solve the problem of poverty—directly—but teachers who are connected to children in a stable professional relationship will be able to help these vulnerable kids better manage the stresses of life, that impact our youngest citizens.

Isaac Asimov said "self-education is the only type of education there is." If this is true, teachers are at best only guides for students who must learn to learn. This effort is only possible if teachers have a trusted long-term relationship with students and their families. This takes years not a semester or two.

Children must feel secure to become effective learners. An innovative solution to building the environment for effective learning is to have teachers assigned a cohort of children—in what is now the first grade—and then let that teacher be that cohort's primary teacher throughout elementary school. The effect of having a professional teacher committed to a class of 25–30 children for 6 years will build bonds between those teachers, those children, their families, and the communities where they live. More important than what grade the child is in, is the impact a teacher can have in helping that child know who they are and that they are capable of greatness.

Some (many) might argue this is social work not teaching. I would respond with a question that all teachers of inner-city African American children must ask themselves: 'Why am I here in this classroom with these children?' I firmly believe that teachers who cannot make this type of commitment, and administrators not willing to support innovation in the education of black children need to consider changing schools or professions.

5.2 You are Either Pumbaa or Timon

September 30, 2021

"Pumbaa: Hey, Timon, ever wonder what those sparkly dots are up there?

Timon: Pumbaa, I don't wonder, I know.

Pumbaa: Oh. What are they?

Timon: They're fireflies. Fireflies that, uh… got stuck up on that big bluish-black thing.

Pumbaa: Oh, gee. I always thought they were balls of gas burning billions of miles away.

Timon: Pumbaa, with you, everything's gas." From Disney's the Lion King.

WEB Dubois said in 1903 that the problem of the twentieth century is the problem of the color line. Because of the COVID-19 pandemic we can update this prescient statement and say that the problem of the twenty-first century is the problem of the rational vs. the irrational. This dichotomy between the rational and the irrational is independent of race, religion, country of origin, gender, or sexual preference. This separation of American society (and the world) into these two camps is not to say that this is all that matters. Traditional human characteristics that divide us, bring us together, and describe us will continue to be important. But right now, those categories and distinctions are minor compared to the hard line that separates the rational from the irrational. We can now ask the question: Are you Pumbaa or are you Timon?

Rationality is defined by a way of thinking that includes logic, reason, and a scientific approach to reality. Irrationality is characterized by impulse, whim, faith, illogic, and wishful thinking. Science and technology are in the realm of the rational.

The scientific method starts with observation. After observation, hypotheses are developed that can be disproved by the data but cannot be proved. Hypotheses are simply if/then statements that show relationships of cause and effect. If you put your hand on a hot stove, you will burn your hand. Scientists take something as axiomatic if the data overwhelmingly supports a hypothesis. We know that the earth is not the center of the solar system. At one time, that was the belief. After data was brought to bear showing the earth circles the sun, most people stopped the belief in an earth-centric solar system. Anyone who continued to believe the earth was the center of the universe was irrational. The irrationals also make hypotheses, but they do not use data to test those hypotheses. The irrational just believe that something is true despite the lack of physical evidence to the contrary.

The current COVID-19 crisis which President Biden once again this week described as a pandemic of the unvaccinated could have just as easily been called the pandemic of the irrational. The dichotomy of the rational and irrational is also revealing itself in other public debates involving Critical Race Theory, election fraud, environmental protection, the federal debt ceiling, foreign and military policy, and the infrastructure bill. It used to be the case that public discussion on important matters was largely a debate among the rational with the irrational delegated to the fringes as "crackpots." Now the debate is between the rational and the irrational. And unfortunately, debates don't work between the rational and the irrational, because debates are based on logic, reason, and facts.

The irrationals do not accept any of these core components of debate. The irrationals storm school board meetings, or meetings that are discussing public health and have nothing substantive to offer but they are pretty good at screaming and crazy talk.

The sooner CT and other states adopt the policies of New York City the better. NYC has told its citizens, if you want to eat in a restaurant, go to a Broadway play, see a movie in a public theater, teach in a public school, or work in a healthcare setting, you must show proof of vaccination. This is a rational policy response to a crisis.

If meaningful debate is not possible between the rational and the irrational—how does society decide what policies are appropriate and effective? I submit that this is not just an intellectual fight that must be won, this is a battle for the future of humanity and the planet. If we listen to the irrational, we will get education that does not educate; healthcare that does not provide health for our citizens; and environmental policy that accelerates climate change. We must confront the seriousness of this dilemma if we are to survive as a democratic society built on a rational understanding of nature and human behavior. We cannot afford to be led by the irrational. There can be no debate.

The strength of anti-democratic forces relies on a large irrational minority of Americans who have demonstrated to a much more sophisticated group of power-hungry manipulators that the irrational will believe anything and do anything if it supports their self-perception, not as irrational but as righteous. The irrationals are being misled for the benefit of those to whom they pledge their allegiance. The manipulators of the irrational who plant the seeds of misinformation know what they say is false.

As rational Americans we must continue to educate and explain to the American people how science works, how democracy works, and how it is a matter of survival that we rigorously and unapologetically seek truth.

5.3 Paying for Ignorance

January 4, 2022

If I decided to sell my SUV and buy a red late model 2-door convertible sports car that has a top speed of 200 miles an hour, my auto insurance company would certainly give me a call to inform me my rates would be higher. It's not like those conservative curmudgeons do not like red sports cars. It is simply that based on their experience, men going through their third midlife crisis who drive red sports cars are at greater risk of accidents.

The same principles hold when you evaluate the cost of health insurance. There is something called an actuarially fair premium that estimates the cost of spreading the total community cost of some activity evenly over the insured population. This concept is particularly important now as we enter the third year of the COVID crisis.

Up to this point in the crisis, all Americans have been paying the health care system for the cost of COVID. I am going to make the case that this is not fair, it is not economically efficient, and most importantly, it does lead to the best public health behaviors and outcomes.

If there were no vaccine that without any doubt lowers the probability of hospitalization and death to near zero, this would not be the case. If the disease hit all Americans randomly, you can make the case that all Americans should pay. But this is not the case with an effective vaccine.

With an effective vaccine we now know several observable facts. One: Over 400,000 Americans will have died from COVID-19 each of the past two years. Two: Since the vaccine, over 95 percent of Americans in ICUs are unvaccinated Americans. Three: There are approximately 100 million unvaccinated Americans. Four: Vaccines are administered at no cost to those who want them. Five: The average cost of one day in an ICU is over $10,000. Six: Forty percent of COVID patients who go into ICUs do not come out alive. Seven: The average stay in an ICU for a COVID patient is 7 days. With this information we can estimate that the total annual cost of ICU care for the unvaccinated is over $70 billion. Therefore, if we were to ask all unvaccinated people to be covered for COVID ICU treatment that policy would cost about $1,000 per person. For a family of

4 unvaccinated Americans, their total COVID insurance premium would be an additional $4,000 on top of their current insurance.

Right now, no such policy is available, but I do not put it past the very aggressive health insurance industry to be working on just such a product. We know that many unvaccinated Americans would probably not be interested in buying such a policy because they are not only unvaccinated, but they also believe they are not going to get the virus. Insurance works best for rational actors.

In the absence of COVID insurance for the unvaccinated, the vaccinated and the unvaccinated are left holding the bill. This is grossly unfair given what we know. It would be like asking all drivers to pay the higher premium on my red two-door sports car (which I do not have now).

An alternative to COVID insurance would be to directly address who has access to the nation's critical care hospital units. There are images from the 1918 pandemic of patients being lined up on cots in large public buildings. Hospital care was at a vastly different place in 1918 than it is today, but a lesson from the last pandemic is that we might want to build out alternative lower-cost facilities for the unvaccinated. We could use closed school buildings, unused warehouses, or other spaces. These spaces would include just palliative care and a limited number of ventilators. The case I am making is that the scarce physician and other critical healthcare human resources should be reserved for those who are behaving in a way that promotes public health. And those that do not behave in that way should bear the consequences of their actions.

These alternative treatment facilities for the unvaccinated should be humane and caring. These facilities can also be open to the misguided pseudo medical treatment wishes of the unvaccinated, who for some reason believe the medicines of quacks are better than the medicines of scientists. Give them all the Ivermectin, hydroxychloroquine, and boiled betal leaves UV light they want. Let their unvaccinated family visit them as much as they want. Let them enter these facilities without masks. Let the unvaccinated nurses and unvaccinated doctors treat them based on their own and similar ignorance. In effect, give them what they want. But keep them out of our hospitals and ICUs. These are not the places for them.

Of course, they could all buy COVID insurance. Or they could all simply get vaccinated.

5.4 Black History Month Is Not Just for Black People

February 5, 2022

In 1835, Alexis de Tocqueville came to America to study our new country to see what lessons could be learned for the benefit of the new French Republic. His book ***Democracy in America*** was an insightful look into the character of America with all its strengths and flaws. A century later, Gunnar Myrdal the Swedish economist and Nobel Laureate gave us perspective of another outsider in his classic 1944 book, ***The American Dilemma: The Negro Problem and Modern Democracy***. Both authors and scholars pointed out the leading role of the relationship between dominant Whites and exploited Blacks in the political, economic, and social life of the country.

I wonder what an outside scholar would think if they saw a former President calling the Black District Attorneys in Atlanta and New York racist because they are prosecuting his criminality. "Madness in great ones must not go unwatched" (Hamlet). What would that scholar think if they were witness to governors from several states promoting laws that ban historical truths about how Blacks were treated during slavery, Jim Crow and today. How would they interpret the bomb threats at 20 historical black colleges and universities? What would they say about the highest court in the land about to overturn affirmative action programs in other citadels of higher education? How would they explain to the world a Senator from Mississippi's claim that the nomination of a Black woman to the Supreme Court would lead to an inferior jurist?

G. Carter Woodson, an American scholar and historian created Black History Month in 1926. His goals were modest and important given the sad state of race relations after WW I. Dr. Woodson was nearly killed in a race riot by returning White GIs in Washington, DC. Woodson's goal when he created Negro History Week was to share the accomplishments of Black Americans with Black Americans that had been hidden by White educators and the media. Everyday Black Americans needed to witness their humanity in a society that was hell bent to deny it and destroy it. While Dr. Woodson might have hoped that all Americans would share in this celebration of Black accomplishment and pride, the goal was to uplift the spirits of race that was beaten down at every turn and made to believe that they were inferior and undeserving of full participation in the American experiment.

Today feels different than any time in our recent history. Never in my lifetime have I felt the simultaneous support of institutions that recognize my abilities, and at the same time what I can only say feels like hatred and terror. While I did not know her, I can understand why a 30-year-old Black beauty queen with a law degree, a TV show, and so much to give would throw herself out of a 30th-floor window. It did not surprise me that another beautiful Black woman would die in the presence of an older White man, and authorities would not investigate until the righteous outrage of the family and a community forced the City of Bridgeport to act. Being Black today, regardless of your degrees, your income, your connections, your personality, your looks, or your neighborhood is a burden that is too heavy for some to carry.

Black history month should be celebrated across this country this month. It is an opportunity to celebrate and learn the familiar and dramatic stories of Crispus Attucks, Frederick Douglass, Booker T. Washington, Sojourner Truth, Bessie Smith, Nat Turner, Jackie Robinson, Mary McLeod Bethune, Josh Gibson, Miles Davis, Jimi Hendricks, Malcolm X, Fannie Lou Hammer, Maya Angelou, Madam CJ Walker, Ralph Bunche, Edward Brooke, WEB Dubois, Hiram Revels, James Baldwin, Muhammad Ali, Martin Luther King, Ralph Bunche, Fred Hampton, and thousands of other notable Black Americans who courageously overcame the odds and made a mark on the sands of history.

I am concerned for the mental health and the continued development of Black children today who in some towns, cities, and states are told only the stories about their past that make White people laugh and smile, but do not bring tears to their eyes. America needs a good cry. Are DeSantis or Younkin, or the local school boards that ban books afraid of the truth, of history, of the ability of their children to learn, or to have a basis to develop empathy? Do they think that White children are incapable of distinguishing a pablum of lies from the truth plain to their eyes? And while Black children need to know their history, so too do White children need to know the history of Black people in this country—not the diluted history, the real history.

In the 1960s movie, ***Inherit the Wind*** based on the 1925 Scopes trial, the fictional town of Hillsboro, Tennessee fired and jailed a local high school teacher for teaching his students the theory of evolution. An out-of-town newspaper man covering the case wrote, "I don't know whether the town leaders had a hole in their heads or their heads in a hole." It looks like DeSantis and Younkin have holes in their heads and want the people of their states to put their heads in a hole. Black History, like the teaching of evolution, or the teaching of the Holocaust, is tough but necessary material for a mature society, and even the children of that society need to learn it. It is ironic that those who would ban books and censor history in their effort to protect, make us less safe, less democratic, and less human. This is a struggle that must be fought and must be won.

5.5 Ignorance and the Three Evils According to Martin Luther King

April 18, 2022

Last Monday, April 4, was the 54th anniversary of the assassination of the Rev. Martin Luther King Jr. That April day in 1968, I was an eighth-grade student at Paul Jr. High School in Northwest Washington, DC.

It started like any other day. We did not have yellow school buses in DC, so I took the No. 70 public bus down Georgia Avenue and got off on the corner of Georgia and Missouri Avenue. (Naming streets after states is a wonderful way to learn the states.) But soon after school started, our principal got on the public address system and curtly announced that school was canceled and told all of us to go directly home.

I went to the bus stop and could see military vehicles driving down Georgia Avenue. The National Guard seemed to come out of thin air. A walker passed and told us the buses were not running. We could see smoke beginning to fill the sky south of where we were.

You can take Georgia Avenue past Howard University to downtown DC and the federal buildings. So, I started my 2-mile walk home with my principal's message ringing in my ears admonishing us to go straight home. It was not until I got home that I learned why the early release, why the buses had stopped, why the smoke in the air, why the military vehicles and why the National Guard. Dr. King had been assassinated.

Exactly a year before his assassination, Dr. King spoke in New York City about the three evils of poverty, racism, and war in a speech called "Beyond Vietnam." Dr. King had long been a vocal and insightful proponent of anti-racism and anti-poverty. But it was Dr. King's criticism of the Vietnam War as immoral that led his critics to conclude he was un-American and others to wonder if his murder was not the work of one man.

Dr. King's "Three Evils" speech is as relevant today as it was 55 years ago. We are approaching the two-year anniversary of George Floyd's public murder. Congress let expire the child tax credit that brought almost 5 million American children out of poverty. And internationally, Russia reminds us how capable humanity is at mass cruelty and what economists call the dead-weight cost of war. Earlier this week, we witnessed random and senseless gun violence in a New York City subway and another police killing of a Black man in Grand Rapids, Mich., after a traffic stop. What is common in all these actions is ignorance.

Racism is an expression of ignorance combined with irrational fear. Some racists are illiterate backwoods rednecks, and some sit in positions of leadership in the public and private sectors. Some racists bristle at the accusation that they are racist and counterattack by saying the accusers are racist themselves or the accusers want to "cancel" them for legitimate views. It is for this reason that intellectual debate on the subject fails.

Racists are ignorant of the humanity they share with people who do not look like them, act like them or share their beliefs. Racists are ignorant about the reality that difference and commonality coexist. We may be different, but we are fundamentally the same.

The Russian invasion of Ukraine demonstrates that government-sponsored murder is evil and relies on ignorance just as racism does. The Russian people (and even their soldiers) are kept ignorant by a regime that knows that if the people knew and understood what their government was doing in their name that government could not stand. Russians who try to enlighten their fellow citizens are arrested. In effect, knowledge is illegal where ignorance rules.

Poverty, the third evil, also depends on the ignorance of the poor, who are a considerable number in almost any democracy but are told that suffering is ennobling and that they will receive their reward in another life. And poverty is considered by some wealthy elites as necessary and unavoidable. But as Vaughn Benjamin of the reggae band, Midnite sings, "Jah make enough for everyone."

Ignorance prevents a reordering of societal priorities that favor war over peace, avarice over economic security and racial animus over tolerance. There is one party that, it can be said, is the party of ignorance. Many Republicans do not want American history taught; they would prefer ignorance. Many do not want to have conversations about non-binary gender identity; they would prefer ignorance. Many do not respect a woman's right to choose; they would prefer ignorance. Many deny science; they prefer ignorance. Many want to restrict voting and illegally manipulate elections; they would prefer ignorance. Many leaders in the Republican Party expressed a preference for Putin over Obama! We cannot afford ignorance.

We can speculate about how the Rev. Martin Luther King Jr. would interpret the events of today. I believe he would be frustrated, but would still have the stony optimism that racism, war, poverty, and ignorance can be reduced through nonviolent confrontation. On the issue of Ukraine, I believe King would not object to the Ukrainians defending themselves against aggression. I think he would see non-violence as the tool to change Russia so that the aggression would stop.

But this will not happen if ignorance prevails. I think King would organize communities in America to understand that protecting our children from unnecessary poverty costs less than the consequences of letting those children suffer in poverty. And King would know that racism needs to be confronted with nonviolent action combined with specific policies to eliminate barriers and racist policies and practices.

And finally, Dr. King would realize that ignorance also exists among the oppressed. The senseless violence plaguing our community is also based on ignorance. Ignorance of others, but most importantly ignorance of self.

April 1968 or April 2022, there is still much work that we must do to confront racism, poverty, war and ignorance.

5.6 Book Banning an Educated Citizenry and Democracy

February 1, 2023

"Your grandchildren won't even understand what a book is." Ray Bradbury Fahrenheit 451.

Joseph Goebbels, the head of the Reich Ministry of Public Enlightenment and Propaganda stated in 1933 "The national education of the German people will be placed in my hands." Not too long after that statement, Nazi student organizations were burning books by Jewish and left leaning intellectuals and writers. Ernest Heminway's books were burned along with the books by Hannah Arendt, Jack London, H.G. Wells, Karl Marx, Baruch Spinoza, and Sigmund Freud. These were all writers whose works are now considered classic works of Western intellectualism. By 1938, Nazi students were burning those books and by 1940, Nazis were committing mass murder on an industrial scale. History shows that autocrats and fascists ban books then burn books, then engage in genocide. This was true in Germany, Cambodia, and Rwanda.

Governor Ron Desantis of Florida has become the self-appointed Minister of Propaganda where his target is diversity, and more specifically literature by and about the Black American experience and the horrors committed by past generations of White Americans. Recently, Governor Desantis authorized and supported the "Stop Woke Act" (HB7) which makes it unlawful for employers to sponsor mandatory employee trainings that promote certain concepts related to discrimination, Diversity, Equity, and Inclusion. This is something former President Trump attempted as well by banning diversity training for federal employees and employers with federal contracts. One of President Biden's first executive orders was to negate Trump's executive order on the topic.

The latest actions by Governor DeSantis to eliminate an Advanced Placement course for African American history and making it potentially illegal for teachers and librarians to have some banned books on their shelves represents the first step in fascism. While difficult to comprehend the reasoning behind DeSantis's bans and animosity to history and literature, it is playing well to certain segments of the Floridian population—other anti-intellectual rednecks who are still fighting the Civil War and the rights that White Americans had to exploit Black Americans.

It is not a coincidence that during slavery, slaves were prohibited from being taught to read. The only occasional exception was that some slaveholders found it useful to allow some slaves to read so they could communicate selected portions of the Bible that supported the "peculiar institution." It was acceptable for slave ministers to preach the parts of the Bible that reinforced the "obey your master" sections but not the "righteous struggle for freedom" messages found in the same book. Despite the restrictions of teaching slaves to read—it was punishable by death to teach a slave to read—blacks read, and they used their human intelligence to see that their condition was not of God, or even of the Bible. Enslaved Blacks learned to root their resistance on Moses' resistance to slavery under Pharoh.

The damage DeSantis is doing to Florida and students in Florida can have lifelong societal impacts. I have zero interest in going to Florida so long as he is governor. I do not think I am alone in this decision. I am a potential retiree who would normally consider the warmth and lower taxes attractive enough to relocate, but not with him in power. But more important than my personal views, the children, Black, White, Asian, and Hispanic, who will get only a portion of the American story from educators restricted by these fascist laws. As a former educator at some of the best universities in the country, Floridian high school graduates will be judged as less prepared for the rigors of higher education. Floridian graduates will also suffer in the eyes of national employers who understand that having a well-rounded education makes for a more productive and effective worker and leader.

The actions of DeSantis are relevant to educators, school boards, students, and politicians here in Connecticut and elsewhere because his playbook is being seriously considered by "conservative" leaning towns and states. It is time for intelligent citizens to look fascism in the face and say "no." If we are to survive as a multi-racial, multi-ethnic democracy we must have a fully educated population.

DeSantis and his ilk are fearful people. Perhaps they had some trauma in their lives related to the people who they now want to "cancel." Trauma is part of the human condition. Fearful people like DeSantis should never be allowed to lead. Fearful insecure people like DeSantis (and Trump and Senator Eugene McCarthy) demonstrate they have not learned from their trauma to be more human. DeSantis has learned that his success is predicated on his ability to make others fearful and for them to blame their trauma and fear on Blacks, Hispanics, Asians, Muslims, and Jews.

In DeSantis' world there would be no Baldwin, no Giavanni, no Wright, no Ellison, no Hurston, no Walker, no Morrison, and no McKinney who had the audacity to call out DeSantis for what he is and his threat to democracy and freedom.

5.7 The Need for a National Black Academy

July 6, 2023

"If you can't join them—beat them." Wu Tang Clan.

Last week, I published an article voicing my opinion on the Supreme Court's decision in the Harvard and University of North Carolina affirmative action case. I admitted that the article was written in a state of controlled rage. I have had time to reflect on the decision and the commentary I have heard from you and others, and have come to a preliminary conclusion that while the Supreme Court's majority does not have the interest of Black Americans as a primary concern, they may have unwittingly done Black Americans a favor.

It has never been clearer that the fate of Black Americans rests primarily in the hands of Black Americans; not the Federal Government, not the Supreme Court, and not even our many allies in White, Asian, Hispanic, and Native American communities. Black Americans should not burn bridges, and we should collaborate when and where we can to obtain mutual objectives, but we should always get something in return for our support. This proposal is not an attempt to re-segregate or cast aside well-developed alliances to the wind. Our true allies will support what I am proposing.

As a child growing up in Washington, DC, my mother realized that I needed help with Algebra and pre-calculus. She hired a tutor who explained Pythagorean geometry better than my teachers did at Paul Jr. High School in Northwest Washington. I was fortunate that my family had the resources to supplement my public education with that type of help from a caring Black woman who had retired from teaching math at the famous Paul Lawrence Dunbar High School. She was not paid a lot of money, and we did not meet for more than an hour once a week. But in that hour, not only did we discuss math problems and principles, but she also instilled in me that math problems were solvable. This knowledge helped me most when I was at Yale Graduate School of Economics in the late 1970s where I had to learn higher forms of calculus and apply that knowledge to social and economic problems.

The economic and social problems Black Americans continue to face will not be addressed adequately by federal, state, or local government. It is time to take matters into our own hands. And first and foremost, we need to establish Black Academies in every city where there are significant Black populations. These academies can and probably should be virtual, the teaching, and mentoring should be voluntary, and they are not in lieu of public education.

At the end of the Civil War, General Oliver Otis Howard led the Freedman's Bureau and was instrumental in the development of today's HBCUs. Howard University is named after General Howard, who was a White ally. At that time, there were no Black Ph.D.s or Black economists, or Black physicists, Black physicians, or Black engineers, or Black millionaires or Black billionaires. The same cannot be said of the situation today.

The Black Academy would fill the gaps that exist today between what our children are learning in public schools and what they need to learn to be independent, successful, and productive citizens of this nation while building up Black communities across the country. The Black Academy would also provide the historical context of Black excellence going back to our African roots. The Black Academy would teach about the Kemet, the Black Pharaohs, the Dogon astronomers, the Zimbabwe walls, and African metaphysics that predates the Jewish and Christian eras by millennia. No one else can do that for us but us.

Is it too much to ask today's Black professionals to offer one hour of their time each week to share their wisdom with our Black youth, who no fault of their own were born into circumstances that have not supported their personal development?

There are many organizations today involved in mentoring. The Links, the Boule, the Black fraternities and sororities, community-based nonprofits and others are doing mentoring today. Now is the time to organize these disparate activities into the Black Academy.

Can White Americans serve as teachers in the Black Academy? Absolutely, if they are allies and buy into the goals and objectives of the Black Academy. Can non-Black children attend the Black Academy. Absolutely, if they are willing to learn and share their resources with the effort. But make no mistake Blacks must control the Black Academy even if that means receiving no support from the federal, state, or local governments. Freedom comes with a price.

Everything cost money. How will the Black Academy be funded? First, all "teachers/elders" are volunteers. Because the Black Academy is virtual, the primary costs will be internet access and a computer or phone. This is a solvable problem.

What will be the curriculum? The curriculum will be based on the life experiences of the teachers/elders. If you are an economist, you can teach economics. If you are an electrician, you teach electrical science. If you are a nurse, you can teach nursing. If you are an investment banker, you teach how to build wealth. If you are an artist, you teach art. If you are an entrepreneur, you teach entrepreneurship. There are external excellent sources that any knowledgeable professional can rely on to support their work with students in the Black Academy.

Is the Black Academy free? Yes, the Black Academy should be free to all students.

What is the result of the Black Academy? It might take two or three generations, but the Black Academy will ensure our children are prepared for the world. The Black Academy will minimize the flow of Black men in the school to prison pipeline and build a Black Academy to success pipeline. The Black Academy will build solidarity within communities. The Black Academy will stress non-violence in our communities and self-policing. It is ignorance of self and history that is the reason Black men are killing other Black men and innocent bystanders. If they attended the Black Academy, this would stop.

CHAPTER 6

Criminal Justice

Abstract Chapter 6 covers the intersection of the criminal justice system and its impact on the racial distribution of income and wealth. The author offers a preliminary theory of violence and the relationships between crime and economic outcomes.

Keywords Crime · Criminal Justice · Violence · Class · Police · Race · Guns

6.1 A Dream Deferred

April 16, 2021

What happens to a dream deferred?

Does it dry up like a raisin the sun?

Or fester like a sore—and then run?

Does it stink like rotten meat:

Or crust and sugar over –like a syrupy sweet?

Maybe it just sags like a heavy load

F. McKinney, *Critical Issues Facing America*,
https://doi.org/10.1007/978-3-031-93217-5_6

Or does it explode?

Langston Hughes.

As a social scientist, I look at data, or in other words I look at history, and ask; are there lessons in that data that help us understand today and make predictions about the future? History tells us to be prepared to be disappointed in the Derek Chauvin verdict. (Chauvin is on trial for the murder of George Floyd.) I hope I am wrong.

Last year was a difficult year in America as we approach the anniversary of the murder of George Floyd. The jury in that trial will be given the difficult task of judging whether former Minneapolis police officer Derek Chauvin is guilty of second-degree and or third-degree murder and other lesser charges. I cannot help but think of the Rodney King case in Los Angeles where a jury found the police officers innocent in the beating that was also captured on video. After the King verdict, Los Angeles erupted in a paroxysm of looting and random violence. I hope I am wrong, but if the jury in Minneapolis finds Chauvin innocent of the more serious charges of murder, a violent reaction in Minneapolis and in other communities is possible, unless communities around the country begin preparing now.

The decision in the Chauvin case has garnered even more attention because of the police murder of Daunte Wright last week in the Minneapolis suburb of Brooklyn Center. It looks like this latest shooting was a fatal accident, but no matter the intention or circumstances, another Black man is dead at the hands of police who are using tactics on our community that seemed to be reserved for Black men. How much can one metropolitan area go through before it explodes? How much can Black communities across the country take before they explode?

The frustration that is taking place in America was intensified this week with the release of a video showing Lt. Caron Nazario, in uniform, being pepper-sprayed by a Windsor Virginia, police officer. In that case the police officer was fired, and the Gov. Ralph Northam has called for an investigation. When Black citizens witness a Black officer of the U.S. Army, in uniform, being abused by police it confirms that no Black man is safe anywhere or at any time.

Currently, 20 percent of all police have military experience compared to 6 percent in the general population. A recent study published in the Journal of Public Health reports that police with military experience are 2.9 times more likely to be involved in a police-related shooting. The

Black community wants police officers who view Blacks as citizens of this country not enemy combatants in Falluja, Mosul, or Kabul.

The resignation and arrest of the officer involved in the Daunte Wright shooting and the firing of the officer in the Lt. Caron Nazario traffic stop will likely not be enough to calm tensions, particularly with the Chauvin verdict about to happen. America is on edge. History suggests that a not guilty verdict in the Chauvin trial could result in widespread unrest across the country. I hope I am wrong.

Black Americans may not have as much income, wealth, or political power as White Americans, but one thing we continue to hope for is equality when it comes to justice. Police shootings of Black Americans is far too common to fit the definition of randomness; this is systematic and requires a systematic solution.

"Defunding the Police" is a slogan, not a policy. The policies that must be changed are addressed in the George Floyd Policing Act, HR 7120. The Act lowers the criminal intent standard, from willful to knowing or reckless. Too often police have been excused for crimes against Black men and women on the grounds that their actions were not "willful."

The Act limits "qualified immunity." Qualified immunity gives police and correctional officers protection against civil suits when they have acted criminally. Police and correction officers currently know that they are personally protected against actions brought by family members of citizens illegally killed. This protection impacts their interaction with Black men. Police officers who act violently are legally bullet proof.

The Act gives the U.S. Department of Justice (USDOJ) the power to act against local police departments who consistently behave in ways contrary to the interests of citizens. The power of the USDOJ to enforce behavior by local police departments gives Black communities added layers of protection from abusive police behavior. A part of this oversight includes maintaining a national database of charges and complaints brought against police officers so that rouge cops fired in one jurisdiction do not just move to another jurisdiction. In the murder of 12-year-old Tamir Rice in Cleveland, the offending officer had been deemed "unfit and emotionally unstable" by his previous employer but was able to get a job in Cleveland PD. This Act would make such moves by criminal and unfit cops difficult, if not impossible.

The George Floyd Act should be passed by large majorities in both houses of Congress. Action must also be taken at the local level. Local elected leaders, business leaders, and community leaders need to engage with black citizens across the country. I would advise elected officials in cities and states across the country to get out of their offices and go to the streets, to the schools, to the churches, and to the businesses where Black citizens live and work. Black citizens need to hear and see elected officials, and more importantly, elected officials need to see and hear Black citizens.

Justice may or may not prevail in the Chauvin murder trial. History suggests it will not, but I am hopeful that it will. Many Americans have literally seen this movie before. I hope this time the ending is different. If this dream is deferred, there might be an explosion.

6.2 The General Theory of Public and Private Violence

May 12, 2023

In three years, economists will be celebrating the 90th anniversary of John Maynard Keynes' treatise, *The General Theory of Employment, Interest, and Money.* The book, written during and inspired by the global economic depression of the 1930s, attempted to explain how major economic stakeholders—investors, workers, banks, consumers, savers, and government—interacted in ways that determined the performance of the macro economy. **The*General Theory*** continues to influence economists and policy makers today.

Keynes demonstrated that understanding how the economy works can be accelerated during times of crisis. Right now, we are experiencing a crisis of public and private violence. A Keynesian approach could be helpful to drive understanding of the underlying conditions that are causing the current paroxysm of violence and random mass shootings that are dominating public consciousness today.

A general theory of public and private violence would start with definitions, terminology, and the destruction of myths. Keynes needed to create new economic terms; a general theory of violence may need new terminology that better captures our current dilemma. Myths about how the economy worked were something Keynes had to dispel before establishing a new way of looking at the economic world. During Keynes' time, the myths he fought were the myths of the classical economists, whose beliefs still resonate with many today. High among those myths was the "invisible hand." A second myth was that the economy would "self-correct."

Classical economists beginning with Adam Smith argued that markets and individual economic actors seeking their own benefit—i.e., profits—created value for everyone else. A corollary of the benefits of the invisible hand is the claimed ubiquity of negative unintended consequences when government interferes with the freedom of market actors.

There are plenty of myths associated with the causes of public and private violence. One of the most pervasive myths is that more guns will result in less violence. This includes tropes like the "good guy with a gun myth"; the "government wants to take our guns to enslave us myth"; the "founding fathers wanted an unregulated armed militia myth"; or the "we are genetically predisposed to be violent" myth. The later myth about humans being predisposed to violence might. like most myths, have some bits of truth, but the data comparing peaceful Costa Rica to its violent neighbors El Salvador and Honduras prove genetics are not determinative. Myths provide comfort to the disillusioned but make no sense.

A general theory of violence must include both public violence (which is violence by the state) and private violence (which is violence among private citizens).

Public violence would also include what we generally call wars between nation states. The United States waged war in Iraq and Afghanistan for over two decades. But before that, there was the Gulf War, and before that the Vietnam War, the Korean War, the two World Wars, the Spanish American War, and, of course, the Civil War. This does not include our incursions in Cuba, Haiti, the Dominican Republic, Nicaragua, Honduras, China, Grenada, Panama, and other places where we sent the Marines, Navy, Air Force, and Army to control resources that did not belong to us and people who were not U.S. citizens. War has a way of metastasizing into problems in the society of even the victors. It is quite predictable that state sanctioned violence against citizens of other countries has a way of blowing back on our citizens. Violence is a contagious virus.

Also, public violence comes as a by-product of policing citizens. "To serve and protect" is how most police departments across the country define their mission. Unfortunately, not all citizens feel served or protected and indeed many of those citizens in low-income and minority communities feel the police are there to protect higher income citizens from their lower-income neighbors. So called "sundown" towns existed where Blacks, Jews, and others were not welcomed as homeowners or by the police who "protected" these White segregated enclaves. The undesirable groups faced arrest if they were not out of those towns by sundown. These practices were common in northern as well as southern states. More recently, the deaths of George Floyd and others at the hands of police led to the international Black Lives Matter movement.

Private violence between citizens is also rampant in the United States with the United States ranked the 16th most violent country among the world's most populous 25 countries. The most violent countries in the world are El Salvador, Honduras, and Venezuela. Is it any wonder why citizens from these countries are lining up at our southern border desperately trying to enter, and why so many U.S. citizens fear their arrival? A general theory would seek to understand why some countries are more violent than others driving their citizens to leave.

A general theory of violence must look at access to guns, economic inequality and poverty, discrimination of all kinds, mental health, international forces, the role of government in the promotion of domestic and international violence, notions of citizenship, education and ignorance, climate change, political stability, economic opportunity, cultural influences, and belief systems.

There are also interdependencies between public violence and private violence as well as domestic violence and international violence. For example, some of our most recent mass shootings have been conducted by men who were trained by the military to kill, only to have those men come home and use those "skills" on their innocent fellow citizens.

Keynes knew, and modern economists know, that complex systems do not lend themselves to simple solutions. A general theory of violence is no different from the complexities of a general theory of macroeconomics. Like all scientific endeavors, we start with observation. We then create testable hypotheses of these complex relationships. Data is then brought to bear to test these hypotheses. Some will be rejected, and others supported. Keynes' work did not eliminate the business cycle, but it did give us a framework for understanding our economic reality so that depressions were never as deep or as long as compared to the business cycles before the General Theory.

The purpose of a general theory of violence is not simply conversational. We must take seriously this inquiry if we are to live in a society where one day the fear of random violent death becomes a thing of the past or at least as rare as it once was.

6.3 Addressing Gun Violence

July 7, 2022

I do not feel safe in America anymore. Long before Highland Park, or Uvalde, or Buffalo, or Virginia Tech, or Aurora, or Pittsburgh, or Newtown, or Charlston, or Parkland, or El Paso, or any of the hundreds of other mass murders of Americans by Americans, it has become difficult to feel safe in public places. These cities and towns have witnessed murders in churches, movie theaters, parades, schools, streets, synagogues, and grocery stores. And these incidents do not include the danger Black men and women feel after George Floyd, Brianna Taylor, and now Jayland Walker in Akron who was shot 60 times by 8 police officers after a traffic stop.

The common factors are guns and social alienation. Social alienation, particularly of young White men when it comes to mass shootings, but also with inner-city youths who feel they have nothing to live for.

It is the availability of guns that separates us from the rest of the world. The fact is, America represents four percent of the global population, but according to Politifacts, we own 42 percent of all privately held guns in the world. There are more than 120 guns for every 100 persons in the U.S. Yemen, a country in the middle of a war is next on this depressing list with about 58 guns per 100 Yeminis. Based on the illogic of the NRA and their Congressional pawns, we should be the safest place on earth, but it is just the opposite.

There are solutions to this crisis that will allow us to feel safe again, but they go beyond the bipartisan bill passed last week and signed by President Biden. The solutions must come from American citizens because too many of our leaders are too afraid of the NRA to lead.

1. The manufacturing, distribution, sale, and possession of automatic and semi-automatic weapons and the ammunition they use, should be illegal. Anyone violating this law should be prosecuted to the full extent of the law. This would include shareholders of manufacturers, retailers, distributors, and private sellers.
2. All persons currently owning these banned weapons should be contacted by registered mail and told to turn in all these weapons to their local police department. We know who legally bought these weapons and we know how to reach them. If these individuals sold or transferred these weapons

to another person, the seller should be made responsible for providing information on the whereabouts of those weapons.
3. To not violate the 2nd Amendment, owners of these turned in weapons will have the ability to join a state approved militia or national guard where their guns will be stored, and training will be provided on their use. If you are not willing to join the national guard, you lose the right to your weapon. The national guard is a "well-regulated militia."
4. Manufacturers, retailers, distributors, and owners of banned weapons are legally responsible for the use of these weapons in any criminal act where these weapons are used. The owners of these companies and individual owners can be sued by victims and survivors of weapons used in crimes.
5. Advertisements of all weapons should be illegal in all media.
6. Gun shows and gun markets should be illegal.
7. Retailers should be prohibited from selling all weapons. Only the State government will be allowed to sell guns to individuals.
8. Possession of all guns in public should be illegal. Cash rewards should be given to citizens providing information about people manufacturing, distributing, or possessing illegal weapons.

Some variations of these laws are necessary for Americans to feel safe again in our homes, in our churches and synagogues, in our movie theaters, in our schools, in our restaurants, in our parks, and in other public spaces.

Government must have a monopoly on guns. This is one market where private markets do not work. In 2021, there were a reported 693 mass shootings which killed 703 people and injured 2,842. So far this year we have experienced 352 mass shootings killing 400 people and injuring 1,300.

It is difficult to value a life, but insurance companies do this all the time. If a life is worth a million dollars, the cost of these violent acts exceeds a billion dollars every year. And this is just the shootings that qualify as "mass shootings." For the parents, the children, the neighbors, the friends, and the co-workers of the victims, no amount of money is enough to compensate them for their loss.

I am afraid if we do not make these or similar radical changes, our future as a democracy is in danger. I am not optimistic about the U.S. Congress implementing anything close to what I am proposing now, but why not Connecticut? The U.S. Supreme Court has indicated in its recent rulings that states have rights that were previously reserved for the federal government. We here in Connecticut should act and show the nation enough is enough. It would be nice to feel safe again in public spaces.

6.4 Jury Duty, Victims, Crime, and Society

January 13, 2023

There are only a few things that are required of U.S. citizens other than obeying the law. You must pay your taxes. If you have children, you must send them to school. If you drive a car, you must get it licensed and insured—something you don't have to do with much more dangerous weapons of war. And you must go to jury duty when called.

My time rolled around to serve on jury duty recently and unlike past times when I was called, this time I had no real excuse for not serving, and this time, I wanted to serve. In the past, as the leader of a not-for-profit organization, there would have been some inconvenience if I had been chosen to serve, but I never got to the point where my inconvenience was a reason for being excused. All the previous times I was excused was because I knew someone, I had some relationship with the company involved in the case, or I had a person close to me who had a similar bad experience as the plaintiff. This time I was hoping to be selected out of a panel of 25 potential jurors. Unfortunately, I was not selected as a juror. I did not even make it to the interview stage, or what is known as "Voir Dire." After 6 hours of waiting, I was dismissed.

Most people who were also not selected were ecstatic. I was not. In fact, I wanted to know why I was not even interviewed. Was it because I have a Ph.D.? Was it because I am a writer? Was it because I am a Black man with dreadlocks and the defendant was also a Black man with dreadlocks?

The rules allow both the prosecuting attorney and the defendant 's attorney to use "preemptory challenges" where the attorneys do not have to provide the judge with a reason for not considering a potential juror. Lawyers for either side can challenge based on cause. I do not know if I was excluded from my civic duty because the defense attorney, or more likely the prosecuting attorney did not like my looks, my education, or my profession. I do know that race, gender, and ethnicity are not supposed to be used as a reason for using preemptory challenges.

Throughout American history, Black and Brown men have been the victims of juries that were not of their peers, and White men have been judged not guilty because Black and Brown jurors were excluded from serving. I am not suggesting that my not being selected is an example of those past abuses. Fortunately, for the defendant, the panel of potential jurors was remarkably diverse in terms of race, ethnicity, and gender, and

I suspect that the six jurors and three alternates were reflective of the diversity of our community.

Before the Voir Dire formally begins, all the panelists (potential jurors) are provided general instructions from the judge and the two opposing attorneys, with the defendant in the room. As I tried concentrating on the instructions from the judge and the introductions of the attorneys, I could not take my eyes off the young defendant. I could not help but contemplate the gravity of the events taking place with him at the center. I could not help but wonder what could have been done differently in this young man's life that could have changed the circumstances and the need for this life-altering trial.

I do not know if the defendant is innocent or guilty, and because I was not selected as a juror, I will only find out if the case shows up in the news. But whatever the outcome of the trial and the decision of the jurors, I cannot help but think about the thousands of other cases where young Black and Brown men are found guilty and sent to prison for their crimes. If you do the crime, you must be prepared to do the time.

Philosophically, we can ask: Is crime part of the natural human condition, or are there systemic causes that lead one citizen to try to take something that belongs to another citizen? This question has been asked for generations by criminologists, sociologists, and common citizens. Reflections on the causes of crime as opposed to debates on how many more jail cells should be built or how long criminals should lose their rights is something that not just "woke" society should consider. The "un-woke" should be concerned because an ounce of prevention is better and less costly than a pound of cure. More jail cells are not the solution. I cannot speak for anybody but myself, but I can distinguish between specific cases of criminality and society's need to address the deeper issues of what we as society can do to reduce criminality. If we do not seriously start addressing root causes, I am afraid we will be fighting a losing battle.

6.5 Crime, Race, Class, and Income

9/13/2023

"I must remind you that starving a child is violence. Neglecting school children is violence. Punishing a mother and her family is violence. Discrimination against a working man is violence. Ghetto housing

***is violence. Ignoring medical need is violence. Contempt of poverty is violence.*"** Martin Luther King.

The Pew Research Center just released a report on its definition of economic class. Pew defines class by income with upper class starting at $156,600 for a family of three; middle class starting at $52,200 and ending at $156,600; and lower class at $52,200 or less.

Defining class purely in terms of income is problematic. One of the enduring features of American mixed capitalist society is the ability for people in the "lower" class to move into the "middle" class and for those in the "middle" class to move into the "upper" class. Like in the Jefferson's sitcom show—the second longest series with primarily an African American cast—the show's theme song, "Moving on Up" reflected this hope and reality. However, we know that class is defined as more than just household income.

A feature of modern American society is for people and families to move in the opposite direction; with people moving from the upper class to the middle class and from the middle class to the lower class. One of the fastest growing homeless populations are baby boomers who were once firmly in the middle class.

Class mobility based on income, however, does not capture a richer more holistic measure of class. A better measure of class should consider one's relationship to societal norms and customs. Criminality is often associated with class. Poor people fill the jails and so there seems to be a direct correlation between class and criminality. But we know that those in the upper classes also commit crimes, but they also do not as frequently end up in prison.

In a report by the Prison Policy Initiative, the pre-prison median incomes for all men, regardless of race, is 52 percent of the median income of non-incarcerated men. For Black men that ratio is 44 percent and for White men the ratio is 54 percent. Poor people do go to jail at higher rates, but does that also mean they commit more crimes? It could be that they are simply arrested more, have access to lower quality attorneys, and are treated differently by juries and judges.

I happened to witness this in action this week in Bridgeport. I was having lunch at a Caribbean vegan restaurant off Main Street in a Black and Hispanic neighborhood. While I was waiting for my meal, two police cars showed up responding to a report that some men were drinking and smoking marijuana outside the restaurant. Fortunately, no one was arrested, but the five or six black men who were questioned by police were upset with the reaction of the police to their activities. Wealthier people can sit outside drinking and smoking marijuana and not be confronted by that behavior for many reasons. Wealthier people are living with more space between their homes and are therefore not as likely to draw the ire of neighbors for their behavior. And secondly, wealthier people are less bothered by the police in their daily lives.

Paying attention to the latest news we know of cases of "upper" class individuals defined by income who behave in ways that demonstrate not only lower-class behavior, but behavior that exhibits no class. Alternatively, we know of low-income neighbors, friends, and family whose behavior is saintly.

Sadly, defining class based on income tends to conflate race and culture with class. Because Black Americans have higher rates of poverty—sloppy thinking, i.e., prejudice, lumps Blacks as low-class even though there are far more White Americans in poverty than Black Americans. The latest census data reports that there are over 25 million White Americans live in poverty compared to 8 million Blacks in poverty. Even though there are three times as many White American is poverty, most people regardless of race do not assume when they see a White American that they are poor or in the terminology of Pew, low-class.

These perceptions have a real impact on policy. West Virgina has a Black population less than 4 percent and only Mississippi has a lower household median income, yet its senior senator, Joe Manchin voted against extending the child tax credit that took millions of Americans and thousands of his citizens out of poverty.

One of the many problems facing poor people today is criminality committed on other poor people. If we are to reduce the current wave of criminality in our cities, we must address the question of why there is a lack of respect for private property and the health and safety of our neighbors. This goes beyond locking up more people. It involves the perceptions that upward mobility remains possible in American society. Without the possibility of upward mobility, the lower class have no rational material reason to behave. Violence begets violence.

6.6 Black-on-Black Crime and Black Community Development

May 29, 2024

Writing about Black-on-Black crime is a difficult subject to cover. I, and other Black social scientists, columnists, and politicians, would prefer to discuss positive contributions rather than dysfunctional aspects of our culture. It is much more comfortable to blame White Americans or the history of slavery, Jim Crowism, discrimination, the courts, the police, and others for the chronic problems in our community. To be clear, there is plenty of blame to go around. However, we must also confront the damage that is the result of our own actions or inaction, even to the extent that our behavior is rooted in, or shaped by, the factors mentioned above.

According to the U.S. Department of Justice, in 2018 there were 2,925 murders of Black Americans and 3,315 murders of White Americans. Eighty-nine percent of murders of Black Americans were committed by other Black Americans. But before making any conclusions, it is important to also note, that while lower, the percentage for White Americans murdering other White Americans is 81%. Statistics on murder and other violent crimes reveal that most crimes, including murder, are crimes of opportunity involving people of proximity and familiarity. Despite the similarity in rates across racial and ethnic groups, Black-on-Black crime has a particularly deleterious effect on the economy in Black communities across the country. While this is an American problem and demands American solutions, Black-on-Black crime also requires a Black solution.

Policing in the Black community has been an ongoing topic of discussion long before George Floyd was killed in Minneapolis. In a 2020 study, after the George Floyd murder, the Brookings Institute reported that between 1997 and 2008, 50 percent of Black males had been arrested and young Black males between ages 20 and 24 were 20 times more likely to be arrested than White males of the same age. As an example, I was arrested twice (with all charges dropped both times) before the age of 21.

The reality is we cannot ameliorate or reverse the problem of crime in the Black community by making excessive arrests, putting a police officer on every corner in every Black community, or doubling the number of jail cells in America. The solution to Black-on-Black crime is building community.

Last week I spoke in Greensboro, N.C., at the International Civil Rights Center and Museum. This museum is at the site of the infamous Woolworth's where, in 1960, four Black students sat at the fountain counter and were refused service. They were then brutally attacked by a mob of White people, none of whom were arrested. At that auspicious and historic location, I talked about the need for Black Americans with resources to move back into low to moderate income Black communities to build a sustainable socioeconomic infrastructure—one that can promote Black businesses, expand job opportunities, and reduce Black-on-Black crime.

Black gentrification could provide the stabilizing force that can build community and develop our own solutions to the problems of business, education, opportunity, healthcare, recreation, and the environment. This is not an easy solution. It is asking Black Americans to demonstrate their racial pride by investing in Black communities, perhaps at the cost of sacrificing some of the comforts of living in White communities. And to our White allies, this is not a rejection of your help and interest in working with us. But we must solve our own problems.

The U.S. Supreme Court recently overturned the South Carolina Supreme Court's decision invalidating a redistricting plan that resulted in 60 percent of Black voters in Charleston County being moved out of their district. This blatant support for gerrymandering by the nation's highest court confirms that Black Americans cannot rely on the courts to protect their political rights; the 1965 Voting Rights Act is effectively a dead letter. The response to losing political control of our communities must be to exert economic and social control. That requires Black gentrification.

By providing economic diversity within Black communities, young Black men and women can see role models. Black businesses could also benefit by having customers with the income to support their businesses. We could finally begin to really educate our children on the rich history of Black people in this country and in antiquity. Black history should be more than one month set aside to commemorate a select few leaders and accomplishments. It should be an ongoing embrace of our rich history and cultural contributions to America and the world. Children who are not taught their historic contributions to civilization do not aspire to match those contributions in their own lives.

Black gentrification and community building would also reduce crime. Gangs would not be allowed to thrive in a Black economically diverse community. Rather, organizations and groups that promote justice, respect, and love would flourish. A reinvigorated Black diverse community could deal with bad behavior without destroying the lives of the men who will one day be responsible for the next generation.

I know this is dreamlike, but we need to dream now more than ever. What has been, is not working. Blacks must be self-sufficient. We might not be able to eliminate Black-on-Black crime, but we can reduce it significantly, and it starts with community.

CHAPTER 7

Musings on Miscellaneous Topics

Abstract This chapter covers a wide range of economic and policy topics that do not neatly fit into the previous chapters, but are important in the lives of Black Americans. But underlying all these topics covered is how there are disparate impacts of policy that may not be intended to harm Black Americans, but unfortunately do.

Keywords Cannabis · Equity · Trade · Housing · Wealth

7.1 The Cannabis Manifesto for Social Equity

April 1, 2022

In late 2021, two large regional not-for-profit organizations, The WorkPlace and CONNCORP came together to form the Alliance for Cannabis Equity for Connecticut (ACECT.org). The purpose of the organization is to support minority entrepreneurs, minority communities, minority workers, and the State of Connecticut's efforts to make the goal of social equity in the new adult-use cannabis industry a reality. Shortly after forming the organization, my firm, BJM Solutions was hired to draft a report on what needed to be done to realize these lofty goals. Earlier this month, ACECT released the manifesto and the associated website (www.acect.org). This is a "good news—bad news" commentary.

F. McKinney, *Critical Issues Facing America*,
https://doi.org/10.1007/978-3-031-93217-5_7

The good news is that the State of Connecticut, through the work of the Social Equity Council and their leadership, is making social equity the law's priority. Many of the states that legalized adult-use cannabis before Connecticut, like our neighbors in Massachusetts only paid attention to social equity after the start of legalization when it became apparent that Black, Hispanic, and lower-income residents did not have a real stake in the business opportunity legalization created. Connecticut learned from these early legalization states that if you do not make social equity a priority in the beginning, it is hard to insert it once the industry gets started. Intentionality is important when it comes to matters of equity. Connecticut deserves credit for this thoughtfulness.

Connecticut's definition of social equity is based on residency and household income. Connecticut residents of any race, gender, nationality, household composition, or criminal record are eligible if they lived in Disproportionately Impacted Areas (DIAs) the first nine of their eighteen years, or five of the last ten years and their household incomes are less than three times the median household income in the state. DIAs are defined as census tracts that are lower income, and experienced a disproportionate amount of marijuana arrests.

Social equity businesses are businesses that meet the first two requirements and at least fifty percent of the business is owned by a social equity entrepreneur(s). The state will allocate licenses to nine different sub-industries, ranging from large-scale cultivators, retailers, shippers, delivery services, and manufacturers. There will be one lottery for social equity businesses and one lottery for all other (non-social) equity businesses. The plan is that the state will have an equal amount of social equity businesses in the industry as non-social equity (traditional) businesses.

Connecticut is about three and half years behind the adult-use cannabis industry in Massachusetts. Massachusetts provides the best guide as to what is possible here. Massachusetts is averaging about $30 million in cannabis retail sales per week. The Massachusetts cannabis industry currently hires over 23,000 workers. Cannabis is the fastest growing industry in Massachusetts and the fastest growing industry in the country. Connecticut has slightly more than half the population of Massachusetts. This means in a few years Connecticut could generate $15 million a week in retail sales and over 11,000 good-paying full-time jobs. State sales taxes on cannabis will total about 20 percent including a 3 percent tax that will go to the municipalities that have cannabis operations located in their towns. The total state tax revenue generated from cannabis could be about $150 million per year. Cannabis will be big business in Connecticut.

The bad news is I am afraid that the supply of legitimate social equity entrepreneurs with the resources and connections is not sufficient to meet the demand. There are similarities in how this industry is being structured and what I experienced in certifying businesses as minority owned and operated. Minority and women front organizations were common when large public and private sector buying organizations first implemented preferential buying programs. These were organizations that put a minority or a woman "out-front" as the face of the business, but the business was controlled by non-minority financial backers. Those non-minority financial backers were willing to assume this role if it meant access to markets that they did not have as a non-minority business.

In a similar way, there will be non-social equity cannabis "investors" who will attempt to either participate in the social equity lottery at a lower cost—social equity businesses pay half of the cost of a license as non-social equity licensees—or skip the lottery entirely. The confusing array of organizational types—social equity partners, equity joint ventures, and Sect. 149 firms—allows entrepreneurs and firms to bypass the lottery, and, or have lower licensing costs. The state's answer to eliminating fronts is to review every social equity applicant and every equity joint venture to ensure that these frauds do not take place. I am not optimistic about the ability of the state at this point to keep these fronts out of the industry, despite their good intentions.

And there still are supply and capacity problems for social equity entrepreneurs. Where are these entrepreneurs going to acquire the capital to pay for licenses, and we have not even started talking about operational costs? Who is going to invest most of the money needed to start these businesses if they are allowed to own only 35 percent of the business? How are these businesses going to be able to assure long-term profitability when the state could control the number of competitors by providing new licenses every year? My fear is that social equity businesses might get established, but in five years, Connecticut's cannabis industry is going to have a minimal number of social equity businesses and minority owners. There are some fixes to these problems, but they will require boldness, innovation, and changes in the cannabis law.

7.2 The Case for Public Cannabis Retailers

May 9, 2022

Connecticut deserves credit for making social equity a cornerstone of the legal adult-use cannabis industry in the State. Other states that legalized before Connecticut had language in their laws that acknowledged how the illegality of cannabis disproportionately and negatively impacted low-income communities of color more than higher income White communities. Connecticut learned that if you want to compensate these communities, there needs to be specific regulations to address these past wrongs by creating real opportunities for these disproportionately impacted parts of our community.

Connecticut's cannabis law creates "social equity" licensees. These are Connecticut residents who live in low-income census tracts today or lived in these census tracks in their youth. Social equity licensees must also have household incomes below three times the state's median household income to qualify.

This past week, the state closed the first round of license applications for lucrative retail licenses. Over 15,000 license applications were submitted. From these applicants, 12 will be provisionally selected. Six of these will be social equity licensees and 6 will be non-social equity licensees. We know that the state allows entrepreneurs to submit multiple applications. Like all lotteries, the more tickets you buy, the greater your chances of winning. We will soon see who wins this lottery.

While it is a lottery with every ticket having an equal chance of being selected, I suspect the laws of probability will prevail and that applicants who submitted and paid for multiple applications will be well represented among the winners.

For the 15,000 losers, they will have another chance in future rounds of retail license loteries. But whether these losers win in the next round or not, we must ask the question: Is turning a handful of entrepreneurs into millionaires the same as social equity because some of those millionaires once lived in low-income census tracts?

The winners do not have to be black or brown. The winners do not have to reside in these low-income census tracts today. The winners do not have to have been arrested for cannabis in the past. There is clearly a disconnect between what we are calling social equity and real social equity that would put money directly in the wallets of people who directly suffered under the regime of illegal cannabis.

It is not too early to consider a different approach to the next round of lotteries to address real social equity. When the Alliance for Cannabis Equity (ACE) completed the Manifesto for Cannabis Equity we focused our attention on the existing law as if there were no major changes that were being contemplated or were politically possible. That was an error.

I am not speaking for ACE, but my proposed change would be for the cities in Connecticut with significant minority populations, Bridgeport, Hartford, New Haven, Waterbury, New London, and others, be allowed to have city owned cannabis retail stores. As an economist, I understand that private enterprises often do a better job at meeting customer needs than public enterprises. But the problems of inefficiency, lack of innovation, and quite frankly public corruption are problems of design. It is possible to run public companies efficiently, innovatively, and without corruption.

If the cities and their taxpayers could own cannabis retail establishments operating within their cities and they could be run like public utilities, there is a chance that this industry could produce the social equity that we claim we want. If the cities owned the retailers in their towns, they could control all the revenue, not some measly three percent of the revenue with their hats in hand. The fact of the matter is, we do not need private companies to control the retail sale of cannabis.

There are examples of this in related industries. If you want to purchase "spirits" in Montgomery County, Maryland, or in several counties in North Carolina, or throughout New Hampshire, you go to the county or the state liquor store. I have been in these stores, and they sell the same products that you can buy in private liquor stores in Connecticut.

The irony of cities selling cannabis to residents who were once arrested by the police in those cities for doing the same thing does not escape me. Yet the case can be made that this is a superior way of spreading the economic benefits of cannabis that is simply not possible by granting licenses to a handful of lucky lottery winners who happen to be social equity applicants.

The criticism of city owned cannabis stores is the uneasiness of some of having their cities directly involved in the cannabis trade. But this is an argument that makes less sense in an era of cannabis legalization. Once cannabis has been legalized, government at all levels is in this business like it is in every other enterprise in the economy once taxes are involved. The cities should therefore seek to maximize their commercial interests while assuring the revenues generated are distributed in ways consistent with social equity.

City owned cannabis retailers could buy from privately licensed cannabis manufacturers, could work with privately licensed delivery services, and have products shipped from privately licensed cannabis truckers. Everything the state does could stay the same. This could include privately licensed cannabis retailers outside of the major cities. Let those private firms compete against city owned cannabis retailers. This private sector competition would force city owned cannabis retailers to be efficient and price competitive. And given the fiscal state of most of our large urban centers, every effort should be made to create and capture income and wealth for our cities. The cities should demand this change for the benefit of all its citizens and to promote real social equity.

7.3 Eating for a Green Planet

April 30, 2023

Last Friday, the day before Earth Day, I was the keynote speaker at the Canadian Aboriginal Minority Supplier Council conference in Toronto. The purpose of the two-day conference was to bring together some of the largest companies in North America and diverse companies to learn how efforts to reduce greenhouse gasses (GHG) can potentially impact their ability to do business with each other. Companies and nations are working to meet carbon reduction goals to prevent global temperatures from increasing by 1.5 degrees. Scientists fear such an increase could lead to extinction of life.

The 2015 Paris Accord required the 196 signatory countries to commit to carbon dioxide (CO_2) reductions—the major cause of global warming—to levels sufficient to save the planet. Canada committed to reduce emissions to 30% below 2004 levels. The United States, the world's largest polluter on a per capita level, pledged to reduce CO_2 to 50% of 2005 levels by 2030.

Reaching these goals will not happen without significant changes in government, business, and individual behavior. Two weeks ago, the Biden Administration ruled that all newly sold vehicles must reduce emissions by 50% of current levels. If successful, that would reduce CO_2 by 10 billion tons. (The largest contributors to greenhouse gases are energy production, transportation, and agriculture.)

In Toronto, I focused my remarks on the actions individuals can take that can help save the planet—and our grandchildren.

Next year will mark my 50th year as a vegetarian. I'm not looking for praise. I will admit the reasons for my decision to forgo the eating of animals and fish was that it was what the cool kids did. But over time, I came to appreciate the health benefits and the immorality of killing animals for food. It was only in the last 25 years that I learned that vegetarianism is the only diet consistent with a green world.

While there is some debate among medical professionals on the health benefits of a vegetarian diet, there are several studies that have linked meat consumption to diseases including colon cancer, hypertension, pneumonia, diabetes, and heart disease, there is little debate about the harmful impact on the environment of a meat-based diet. Cattle farming is one of the largest contributors to GHGs. Additionally, it is meat consumption that caused epidemics, including bird flu, swine flu, Ebola, and COVID.

There are 29 million beef cattle in the United States. They consume 75 trillion gallons of water annually and each one of those cows produces 220 pounds of methane gas, a significant GHG. Added to this is the impact on the land and ground water from the millions of tons of cattle, pig, and chicken waste.

Many of us learned as children that it is not proper to discuss politics, sex, or religion at the dinner table. That taboo has been extended to discussing what's on that dinner table. However, we must have this conversation. I know converting the readers of this column from meat eaters into vegetarians is an unrealistic pipe dream, but meat eaters need to confront the environmental damage they are doing by simply eating a hamburger.

The meat industry is a powerful protector of their economic interest. In 1998, Oprah Winfrey was unsuccessfully sued by the Texas Cattle Association because of comments she made in support of vegetarianism on her popular show.

Every year, Americans produce over 20 tons of CO_2 per person compared to 4 tons for everyone else in the world. Modernity has significant advantages. However, our personal consumption behavior in terms of how we heat and cool our homes, what we drive and how those vehicles are powered, and what we eat can no longer be ignored. We do not have to return to eighteenth century living, but we must change. If we do not, rest assured, Mother Earth will have the final say.

7.4 To Solve the Immigration Problem, We Must First Admit We Created the Problem

June 9, 2023

Fred McKinney, Ph.D.

"PBSuccess was the code name for a covert operation carried out by the U.S. Central Intelligence Agency (CIA) that deposed the democratically elected Guatemalan President Jacobo Árbenz Guzman and ended the Guatemalan Revolution of 1944–1954. Guzman was seen as a serious Communist threat after confiscating two-thirds of United Fruit Co.'s 332,000 acres and legalizing the Communist Party." (BACM Paperless Archives)

With thousands of migrants lining up on our southern border and governors threatening to charge other governors with kidnapping for transporting Central American migrants from their red states to rival blue states, we are so fixated on this serious human, political, and moral crisis that we forget—or do not acknowledge—the historical roots of this current crisis. If we ask why we are experiencing this crisis and we objectively look at the historical facts, we will be forced to acknowledge our own culpability.

The quote above comes from recently declassified CIA documents that clearly show that it was United States policy to destabilize the Central American countries from which these migrants on the border are coming from today. In international affairs, like life in general, karma is real.

The CIA documents reflect U.S. policy that viewed all international relationships from the perspective of the Cold War and fears of Soviet intervention in countries south of our border. The Soviet threat and interest were real, but our reaction to that threat was a mistake that has ongoing consequences today.

Instead of promoting the economic interests of Guatemalans, Hondurans, Salvadorians, Nicaraguans, or Haitians, we sent the Marines and the CIA to destabilize democratically elected governments on the pretense of preventing Soviet influence when we were there, at least in large part, to protect the private business interests of the United Fruit Company and other U.S. based businesses. It is not a coincidence that the sibling heads of the CIA and the U.S. State Department, Alan and John Foster Dulles, were directors of the United Fruit Company. Talk about a conflict of interest.

The gangsters we supported throughout the Caribbean and Central America in the 1950s, 60 s, 70 s, 80 s, and 90 s and into this century are the reason the people from these forlorn lands are heading north. Truillio in the Dominican Republic, Duvalier in Haiti, Samosa in Nicaragua, Batista in Cuba, Velasquez in Honduras, and Martinez in El Salvador are just a handful of the U.S. supported dictators who enriched themselves at the expense of their people and the stability of their countries. We supported these thugs as long as they promised to keep the Soviets out of their countries. And if they did that, we turned a blind eye to anything they did to their own people or their country.

This history is why we have a crisis at our border. If we have any hope of solving this problem, we must acknowledge the damage we did to these societies and be prepared to invest in democratic institutions, land reform, education, and human rights in all of these societies. If we are not prepared to do this, the stream of migrants coming north will not stop.

Many Americans will choose to deny the connection between our actions in Central America and the Caribbean and today's immigration crisis. They may also argue that our policies to destabilize these countries was an appropriate reaction in the Cold War environment. Those who oppose this analysis may also argue that we cannot or should not use the moral standards of today to evaluate events of the past.

Each of these arguments can be easily refuted by the data. The migrants from these countries are coming because there is no security and little opportunity for people who want to simply support their families and live their lives in peace. We disrupted their countries in ways that are difficult for us to imagine. We created the thugocracy that has entrapped millions of our neighbors.

I do not claim to have solutions to this complex problem that has taken decades to create. But I know that punishing the migrants who are the historical victims of our policy is not one of them. I think most Americans want our foreign policy to reflect our democratic ideals. The fact that they did not even come close to those ideals in the past should not prevent us from pivoting and moving forward in a way that will make our children proud of our decisions. Reading the CIA's history of assassinations and political intrigue is a shameful history that we cannot change, but we can correct our actions going forward if we have the courage to base our actions in Central America and the Caribbean on our democratic principles.

7.5 The Impending Great Migration of American Women

May 26, 2023

"Our Negro problem, therefore, is not of the Negro's making. No group in our population is less responsible for its existence. But every group is responsible for its continuance.... Both races need to understand that their rights and duties are mutual and equal and their interests in the common good are identical." Isabel Wilkerson, *The Warmth of Other Suns*

As soon as Black Americans had the ability and resources to leave the Deep South after the Civil War, they left. After Jim Crow laws were enacted, as soon as Union troops were removed from the defeated Confederacy, Black Americans left. Over six million Black Americans moved from the former Confederate states of the Civil War era to the Union states between 1910 and 1970. My wife and I were born in the South and were two of those six million migrants. My wife's family moved from Madison Heights, VA, near Lynchburg, to Newark, NJ, and my family moved from Pine Bluff, AR, to Washington, DC, both in the late 1950s. Both of our families moved to pursue better economic opportunities and less oppression and discrimination.

Jim Crow laws were America's shameful version of apartheid resulting in racial inequality and state sanctioned terror. Jim Crow laws restricted every aspect of life for Black Americans making it impossible for Blacks, or for that matter White Americans, to reach their human potential. But while Whites suffered from the contagious disease of racism, they also benefited at the expense of their Black neighbors.

The same states that practiced the most pernicious forms of Jim Crow are also the states that today restrict the healthcare rights of women. The lesson of the "Great Migration" of Black Americans is that people can and should vote with their feet. Women—by the millions—must be at least contemplating leaving these states and moving to states where their rights are duly respected.

As of this week, 15 states have passed total bans on abortion since the Supreme Court's overturning of the Roe v. Wade decision. These 15 states do not include Georgia, which recently passed a ban after 6 weeks, but they do include, Texas, Mississippi, Alabama, Arkansas, Tennessee, Kentucky, West Virginia, Missouri, Oklahoma, Wisconsin, South Dakota, North Dakota, Idaho, and Nebraska. The female population in these states is approximately 60 million.

Women from these states might not want to attend a university, take a job, start a business, buy a home, teach in a public or private school, pay taxes to a misogynistic government, or raise a family in one of these restrictive states. Corporations know this and will factor their own location decisions on the availability of talent. Disney recently announced the cancelation of a billion-dollar project in Florida because of human rights restrictions by Governor Ron DeSantis. The NAACP, Equality Florida, and the League of Latin American Citizens (LULAC) have issued travel advisory warnings for their constituents considering vacationing in Florida. Former Miami Heat Hall of Fame player, Duane Wade recently announced he and his family were moving out of the state because of the state's treatment of trans children. I might have to travel to Florida for business, but I go reluctantly.

States, like Connecticut, should roll out the welcome mat and encourage the unfortunate women who, by virtue of the lottery of life, were born or currently reside in these restrictive states. They do not have to stay. Many will leave. They can come here and pursue their dreams. We should be taking out ads in restrictive states saying they are welcome here.

The economic impact of a gender-based migration would exacerbate already significant differences in income and wealth between the radical religious states and Connecticut. In Connecticut, we tend to look at the glass as half empty. But consider that the median income in Mississippi is more than $30,000 less than the median income in Connecticut. The lower cost of living in these states does not offset the lower incomes. Life is harder in these states and will only get worse.

The women and men who leave the restrictive states will be those with the greatest drive and skills, which will further depress economic conditions in these states while Connecticut and other recipient states will be richer because of the incoming talent. Yes, we should gladly welcome these women entrepreneurs, teachers, physicians, executives, engineers, nurses, attorneys, scientists, leaders, taxpayers, and most importantly citizens who respect the rights of women.

A gender-based migration would increase the political contrasts between the restrictive states and states like Connecticut in ways reminiscent of how slavery divided the union before and during the Civil War. It is scary to think that human rights could once again be the basis of a political demarcation that is already playing out between our two major political parties. The road to political peace will depend on women's healthcare rights. The obvious solution is for women in the restrictive states and the men who support their rights to use democratic methods to reverse the madness in their home states. If they are not able to do this, or do not think it is possible, we must be prepared for another Great Migration of American women.

7.6 The Only Solution Is a One-State Solution

November 22, 2023

"It gives me great pleasure to announce to Congress that the benevolent policy of the Government, steadily pursued for nearly thirty years, in relation to the removal of the Indians beyond the White settlements is approaching its happy consummation." Andrew Jackson Speech to Congress on the Indian Removal Act

On September 17, 1978, President Jimmy Carter, Israeli Prime Minister Menachem Begin and Egyptian President Anwar Sadat signed the Camp David Accords. This historic agreement brought an end to hostilities between the two warring countries that have stood the test of time. This peace cost Sadat his life. A week before the Camp David Accords were signed, I started my job at the White House as a junior economist on the Council of Economic Advisers. While I had nothing to do with the Accord, it was clear that this was a major accomplishment that could have secured the reelection of Jimmy Carter. However, within a year, on February 19, 1979, the Shah of Iran was overthrown, 52 Americans were captured and held hostage for 444 days, and Carter's presidency was doomed.

President Joe Biden's chances for a second term have been a source of anxiety for many Democrats, even when his likely opponent is a 4-time indicted, twice impeached, ex-president who is widely characterized as a fascist, and who could very well be convicted of federal and state crimes before January 2025.

Parallels between Biden and Carter are relevant now because of how the Israelis are executing the "war" in Gaza. While Israel, like any nation, is entitled to defend itself against terrorist attacks, the scope of its response and tactics employed raise serious humanitarian concerns. To many people around the world, this looks more like a police action that Black communities in this country are all too familiar with, than a war between two armies. It is for this reason that many Black Americans I speak with are deeply troubled by President Biden's unqualified support for what history is sure to say is an ongoing war crime and genocide in both Gaza and the West Bank.

Israel is appealing for international empathy for the loss of over 1400 of its citizens in the atrocious terrorist attack of October 7, 2023. Like many peace-loving people in the Black community and elsewhere, we feel their pain at the slaughter of innocents, and support the families of the dead, the injured, and those taken hostage.

American Jewish thought leaders and columnists are appropriately denouncing the rise in antisemitism throughout American society in schools, universities, on the internet, and in workplaces. I join them in that denunciation of racism and antisemitism. But this does not mean I cannot also denounce Israel for its actions in Gaza and the West Bank since October 7. It raises the question: Can you be anti-Zionist and not anti-Semitic? I believe you can. I spent three years on the faculty at Brandeis, loving every minute of it. I do not think I would have been hired or would have continued to work in this premier Jewish institution of higher learning had I been an Anti-Semite.

The Israeli Palestinian problem in modern times is the direct result of pursuing the impossible—a two-state solution.

The Oslo Accords, signed on September 13, 1993, by President Bill Clinton, Israeli Prime Minister Yitzhak Rabin, and PLO leader Yasir Arafat formed the basis for the two-state solution that was supposed to bring peace between Israel and Palestinians. Thirty years have passed since then, and it is reasonable to conclude that the two-state solution is not working, and never will. Ironically, on this point, I agree with current Israeli Prime Minister Netanyahu, who believes that a two-state solution is not feasible or desirable for Israel. He and his right-wing party are doing everything they can to prevent a Palestinian state from being successful.

Netanyahu and past Israeli administrations have pre-determined the failure of a two-state solution by building Jewish-only settlements in the occupied West Bank, which theoretically is the only place a Palestinian state could exist. There are now over 200 settlements with over 500,000 Israelis living in the occupied West Bank. These should not be called settlements. They are newly constructed, permanent urban mini-cities with separate Jewish- only highways and twenty-foot-high concrete barriers and barbed wire fences that separate the Jews from the Palestinians. To Black Americans, who see this, we cannot help but be reminded of our own Jim Crow experience and the Bantustans in Apartheid South Africa. For Native Americans, this separatism is a reminder of desolate reservations and the Trail of Tears (started by President Andrew Jackson, the first president elected when White men without property could vote for the office),

President Biden, if you are listening, the solution to this crisis is not a two-state solution, but a one-state solution where all people currently living in "greater Israel" become citizens of one country based on equal human and political rights, regardless of religion or ethnicity. A truly democratic Israel with one person—one vote is the only long-term solution to this age-old dilemma. The two-state solution is only a recipe for continued segregation, wars, death, and destruction. President Biden, you have an opportunity to change the narrative in the Middle East by calling for a solution that is unequivocally consistent with our own national principles of democracy and human rights.

I know the argument against real democracy in Israel is demographic. We hear it in this country today as some White Americans fear becoming the minority. But democracy is not and cannot be based on anything other than our humanity—not our race, our ethnicity, our sexual orientation, our religion, or lack thereof. The one-state democratic solution is an American solution and the one that President Biden needs to pursue. We all need to dump the flawed and failed fiction of the two-state solution that has not worked in 30 years and will never work. Israel needs a Lincoln not a Netanyahu.

For Biden, he runs the risk of losing Black, Native American, and young people's support in the 2024 election, not to Donald Trump, but to non-voting—- an outcome that might result in Trump's election and the end to American democracy as we know it today. Biden has a tightrope to walk, but if he wants to be a transformative president, he needs to be on the side of democracy.

7.7 Moral Capital

January 5, 2024

Economists from the days of Adam Smith have recognized the importance of capital in the economic development of nations. It is no surprise that Karl Marx's most famous book on the nature of capitalism is titled "Capital." Capital comes as financial assets, primarily savings that can be converted into physical assets that add to the productivity of labor. Capital, labor, and land are the major factors of production in a capitalist economy.

The Nobel economist, Gary Becker won this distinction because of his research in what is now known as human capital. Human capital represents the knowledge and skills that society, family, and individuals invest in people, making them more productive. Being more productive is the capitalist explanation for why one worker earns more than another worker. Education is viewed by Becker and other economists as a key manifestation of human capital. Studies show on average, American workers with less than a high school diploma earn $682 per week in 2022 compared to $1,432 per week for workers with a bachelor's degree and $2,080 per week for workers with professional or doctorate degrees.

The concept of political capital is used to describe the power of influence one person has over other people with power. Political capital is used to influence policy and events.

Financial capital, human capital, and political capital all have the property that you can make investments that increase the amount of that capital, and the reality that as that capital is used it depletes. Economic textbooks use the metaphor of a bathtub to demonstrate the relationship between investment, depreciation, and the amount of capital. Investment is represented by the water going into the bathtub, the drain represents the depletion of capital because of depreciation, and the level of water in the tub represents the amount of capital.

There is also something related to political capital called moral capital. Moral capital can protect a nation, a company, or an individual from errors of commission and omission that cause harm to other nations, companies, communities, or individuals. It amounts to giving nations or others a mulligan, a pass or a do-over. Like other forms of capital, moral capital is depleted as it is used and can be added to by new moral actions.

Rev. Martin Luther King understood the power of moral capital. He called "undeserved suffering" by nonviolent protestors at the hands of racists police and politicians as adding to the moral capital of the civil rights movement. Dr. King believed nonviolent resistance designed to overcome injustices brought needed media attention to that injustice. Nonviolent resistance also has the transformative power to change the heart of the oppressor.

Moral capital, like all forms of capital, is depleted when it is used or when immoral acts are committed. A good example of the accumulation and depletion of moral capital is what happened after 9/11. After 9/11, the United States had the empathy of the world. Nearly 3,000 people died undeservedly on 9/11. Over the next twenty years, the United

States squandered and depleted that moral capital through the wars in Afghanistan and Iraq. According to the Watson Center at Brown University, over 7,000 U.S. troops and over 8,000 contractors were killed along with 177,000 Iraqi and Afghani "combatants." According to the Brown University study, the total loss of life of these twenty years of war exceeded 4.5 million civilian deaths. I wonder, if we knew this was going to be the outcome before we started this level of massive state sanctioned violence and the hundreds of billions of dollars spent, would we have considered an alternative response to 9/11.

Israel never had universal support for its Zionist experiment, but it did have a great deal well-earned moral capital as result of the holocaust when as many as 6 million Jews suffered from undeserved violence and genocide. Today, the Zionists indiscriminate murder of Palestinian children and other innocents is depleting their moral capital to the point where they are now a pariah around the world. They did not heed the advice of President Biden, who warned Netanyahu to heed the lessons of our mistakes in Iraq and Afghanistan where we depleted our moral capital. Ironically, the moral capital of the United States is also being depleted because of our support for the acts of Israel. This decline in moral capital for the country is also depleting the moral capital of President Biden, Israel's chief defender.

Like other forms of capital, the only solution to declining moral capital is to invest in moral actions. This is different from just mouthing moral words. Like other forms of capital, the only solution to declining moral capital is to invest in moral actions. This requires action. The solution for Israel's depletion of its moral capital is to radically transform its behavior from what is doing now to something that looks more like real democracy and the restoration of Palestinian rights and land. Unfortunately, there are no other options if Israel wishes to return to the international community as anything other than an immoral country.

President Biden has a moral capital problem that will impact his reelection campaign. If there is a case for his blind support and complicity for Israel, he has not successfully made that case. If anything, he has made it worse with the recent decision to send $147 million in new tank weapons without Congressional approval. It's good advice to a man trapped in hole of his own making to stop digging. I do not know who is advising the President, but his position on Israel and their attempt to extinguish the Palestinian people will cost him millions of voters who supported him in 2020 because he is depleting his moral capital.

It is not too late for President Biden to replenish his moral capital through actions that lead to peace and justice in Israel. Moral actions often take more courage than violent actions. America special place in the world is grounded in our moral capital.

7.8 It Is Time to Re-Establish Normal Relationships With Cuba

January 18, 2024

I was in the third grade at Shepherd Park Elementary School in Washington, DC, in October of 1962 when the Cuban Missile Crisis happened. I remember it clearly because that month we began practicing "duck and cover" air raid drills every day. In these drills, we ducked under our wooden desks and were told to stay there until the all-clear alarm had been sounded. For an 8-year-old, I had no idea how close we as a nation were to nuclear annihilation.

That was over 60 years ago. For most of that time, with a brief period when President Barack Obama and Cuban President Raul Castro with the assistance of Pope Francis negotiated a deal to begin to restore full diplomatic relationships with Cuba, the United States has imposed a devastating and destructive embargo on the Cuban people. One of the first things former President Donald Trump did was throw a bone to the Cuban exile community in South Florida by reversing the Obama 2014 Executive Order. During the time of the Obama Executive Order, Cuban-Americans could remit money to their families in Cuba. Americans could vacation in Cuba. And American investors could begin making investments in the largest Caribbean Island that before the 1959 Cuban Revolution had a per capita income that was greater than the per capita income of Ireland or New Zealand.

Our Bay of Pigs fiasco, the following embargo, and the missile crisis began a long downward slide in the Cuban economy and Cuban society. This was compounded when the Soviet Union collapsed in 1989 leaving Cuba without its primary financial lifeline.

Following these events, the weakness of a planned economy became glaringly clear. Production of agricultural products, including sugar, declined. Long lines and inefficiencies were everywhere. Black markets developed to fill the gap but were insufficient to meet the needs of the people. Corruption of government officials and ordinary Cubans was rampant. Cubans got on rickety rafts and attempted to cross the 90-mile journey to U.S. shores. Many Cubans died attempting the dangerous crossing.

One of the biggest debates taking place right now is regarding our immigration policy. I have yet to hear anyone make the connection between the Trump restoration of the embargo, and President Joe Biden's decision not to return to the Obama-era policy and the exponential growth of Cuban migrants at the United States southern border.

According to the Washington Office on Latin America, the number of Cuban nationals encountered at the southern border has increased from 13,410 in 2020 to a staggering 220,908 in 2022. Several factors have led to this increase, but the return of the embargo and resulting economic consequences of these restrictions is one of the primary factors. After Trump reinstated the embargo, 400 offices of Western Union, the leading firm processing remittances from the United States to Cuba, shut their doors in Cuba. Remittances were a major source of foreign exchange for Cubans allowing them to purchase much needed goods and services in markets where the Cuban peso could not be used.

We made our points. Castro was bad. Planned economies do not work. Russian missiles 90 miles from our border will not be allowed. It is time for the Biden Administration and Congress to normalize relations with Cuba, not just for the economic stability and well-being of Cubans in Cuba, but because it is in our national interest as the immigration numbers continue to grow. A quarter of a million Cubans would not be crossing the Mexican border if they had opportunities to work and thrive in their own country.

I understand why Biden has not reinstated the Obama-era Cuban policy on U.S. domestic political grounds. He has extraordinarily little chance of convincing the generation of Cuban-Americans who came here in the 1960s to vote for him. His tacit support for the embargo will not help him with this group of voters. They are dyed-in-the-wool "anti-communists" determined to destroy the Cuban people on the island with our assistance.

Normalization with Cuba would bring capital to a society that is crumbling into the ocean like the 1956 Chevrolets that serve as taxis on the Malecon. American tourist dollars could help transform this economy to what it once was—the jewel of the Caribbean.

It is unlikely President Biden would lift the embargo before the election, but if he is reelected and Congress does not want to make the normalization permanent, he should issue his own executive order normalizing relationships with Cuba.

The United States consumer and investor can have a bigger impact on internal Cuban politics than our failed 60-year embargo policy has achieved. If we can agree that democracy and human rights are good for Cubans, we should be able to agree it is time to normalize relationships with Cuba.

7.9 Real Estate Investment Trust Are Not the Source of the Problem Facing America's Safety Net Hospitals

March 15, 2024

It is easy for people to be confused about matters of cause and effect. It is only through understanding how things work that we can distinguish between the two. For example, to an outside observer looking at a gas-powered car, it would be easy to conclude that the wheels are the source of mobility. Their movement is readily visible, and as they move, the car moves. We know, however, that the engine is the source of the power, and the movement of the wheels is the result. Cause and effect.

Several states including Connecticut, Rhode Island, Massachusetts, Pennsylvania, and the federal government are considering legislation that would restrict the activities of private equity companies and Real Estate Investment Trusts (REITs) in the hospital industry. The motivation for this legislation includes growing worries about the financial health of hospitals, particularly hospitals serving low-income and rural communities. These safety-net institutions are experiencing reductions in both patient care quality and access while generating huge dividend payments to private equity company owners. Several for-profit hospital chains in these states have shut down or threatened to shut down while the private equity firms have made millions. This, however, begs the question about the cause of the hospitals' faltering economic conditions.

In Connecticut, Prospect Medical Holdings owns the three financially troubled Eastern Connecticut Health Network (ECHN) hospitals in Rockville, Waterbury, and Manchester. Several Steward Hospitals in Massachusetts are teetering on bankruptcy. ECHN and Steward are owned by private equity companies, and both hospital systems sold their real estate to a hospital REIT.

The experiences of ECHN and Steward are not unique. Between 2010 and 2021, 140 rural hospitals throughout the United States closed their doors. And hospitals in cities including Chicago, San Antonio, Seattle, and Los Angeles have gone out of business.

While there are legitimate questions about private equity, hospital REITs such as Medical Properties Trust (MPT) provide significant value to troubled institutions. Hospital REITs buy the real estate of hospitals and then lease the facilities and land back to the hospital in long-term leasing arrangements. Hospitals get hundreds of millions in cash, which they can use (and in some cases must use) to improve the facilities, hire needed staff, and improve operations and quality. The REITs get a long-term lease with the hospital operator, but do not operate the hospital. Lease payments to REITs average less than 6 percent of hospital revenues. Many corporations outside of health care now have REITs as their landlords for buildings they once owned.

The growth of private equity acquisitions and sale leasebacks are the result of the poor economics facing safety-net hospitals—not the other way around. There are more than 6,000 acute care hospitals in the United States. They are in rural areas, urban areas, suburban areas, Native American reservations, and military facilities. Hospitals can be for-profit organizations or not-for-profit organizations. They may be secular or religious in their origins. Hospitals are challenging organizations to manage. Their fortunes depend on third-party reimbursement, a complex regulatory infrastructure, uncertain demand, rapid advances in technology, a plethora of licensed professionals, and deep ties with public and private sector leaders. Not all hospitals survive these challenges.

On average, if a hospital receives a dollar for a service from a private health insurance company, they will receive $0.60 from Medicare, $0.40 from Medicaid, and $0 from uninsured low-income clients for that same service. Uncompensated care can represent 20 percent of the care delivered in these "safety-net" institutions, not because that is the market they are strategically pursuing, but because of where the hospitals are located.

Urban and rural hospitals are closing and have been closing in communities because of their patient-mix. Without substantial philanthropic contributions, safety-net hospitals cannot cover their operational deficits, eventually leading to bankruptcy.

Hospital REITs offer hospitals a lifeline by providing them with much needed cash. Hospital REITS own tens of thousands of hospital beds and hundreds of hospitals nationwide. Only a fraction of those hospitals cannot pay their rent. Sale leaseback transactions benefit both the buyer and the seller, otherwise they would not happen. In the absence of other sources of capital, REITs are their best option.

There is no question that America's safety-net hospitals are in trouble. But it is important to understand this trouble started before private equity companies and hospital REITs entered the market. Congress, state legislators, and others need to evaluate the legal environment for private equity in hospital acquisitions. However, there is no question that the financial problems facing safety-net hospitals are not the result of sale leaseback transactions. The problem is reimbursement. If we want urban and rural safety-net hospitals to survive and continue to provide services to low-income and undocumented residents, we need to pay hospitals appropriately for their services. Attributing blame to hospital REITs for the problems of safety-net hospitals is illogical, lacks an understanding of hospital economics, and is wrong. Any legislation in this area must discern the difference between cause and effect. If it does not, the legislation will make things worse.

7.10 Understanding the Current Haitian Crisis—This Is a Revolution

April 10, 2024

"The Problem for Haiti is that God is too far, and the United States is too close." Dante Bellegarde, the Haitian representative to the League of Nations.

Last Sunday, April 7th, marked the 222nd anniversary of the death of the great Haitian leader Toussaint L'Overture. He died like a pauper in a cold damp French prison in solitary confinement. This remarkable man led the most successful slave revolution in the history of humankind! The Haitian revolution started in 1791, just two years after the French Revolution in 1789, and only 15 years after the start of the American Revolution in 1776. The Haitian Revolution influenced the entire Western world for the next century. Most Americans do not know that without Toussaint L'Overture's victory over French, English, and Spanish colonizers in Haiti, we might not have an America from sea to shining sea. The loss of Haiti made the Louisiana Territory worthless to the French and available for sale to the Americans in 1803.

Ironically, it is with both great sadness and hope that I watch the current crisis in Haiti and how it harkens back to the great Haitian revolutionary leaders like L'Overture, Dutty Boukman, Cecile Fatiman, Jean Jacques Dessalines, and Henry Christophe. Today's problems are firmly rooted in this past.

In 1825, with the looming threat of French re-enslavement on the island, Haiti agreed to compensate France for French recognition of the country and to "repay" the Haitian slaveholders for the loss of their "slave property" after slavery was abolished. Haitian payments to France did not stop until 1947.

In 2003, President Jean Bertrand Aristide demanded that the French repay these funds. In today's dollars, that debt amounted to $30 billion. In response to that demand, the French, the Americans, the Canadians, the Vatican, and the Dominicans overthrew Aristide on February 29, 2004. This 2004 coup d'etat, or what some have called a kidnapping, was the second time Aristide had been forcibly removed from office after being democratically elected.

In the 1990 Haitian election, Aristide won 67 percent of the vote. In 2000, he won an astonishing 91 percent of the vote. There is no doubt that Aristide and his Lavalas party had the overwhelming support of the Haitian population. But he did not have the support of America, France, Canada, the Dominican Republic, the Vatican and others in the international community who theoretically supported popular democracy but opposed the righting of historical wrongs.

The violence in the streets of Port au Prince today has been labeled by Western media as "gang violence" and anarchy. The release of political prisoners is a pretext for why "we" need to intervene. The Western narrative is that the violence is random, unnecessary, brutal, and uncivilized. The images repeatedly shown are designed to make the public equate the Haitian people with barbarism—unlike us. This makes intervention by the United States and our allies even more justifiable.

The United States led blockade during the second Aristide administration cut off all economic aid from the World Bank, the IMF, and the Inter American Bank. This, combined with France's refusal to make any reparations for the money owed to the Haitian state, led to the destruction of the Haitian infrastructure and public services, and resulted in the worst poverty in the Western Hemisphere.

On July 7, 2021, 28 Colombian mercenaries, trained, billeted, and protected by the Dominican Republic and funded by the United States, assassinated Haitian President Jovenel Moise. This was the latest example of imperial intrigue that has plagued Haiti since its founding in 1804. And now, the United States and France are attempting to pay the Kenyan government to outsource police forces to Haiti for $200 million to suppress what is nothing less than a revolution. I guess four occupations of Haiti by U.S. Marines is enough.

This is a revolution, not gang violence. The United States and all foreign powers need to stay out of Haiti unless they are prepared to support the interest of the Haitian people who are overwhelmingly poor. Trickle-down economics does not work in the United States. What makes policy makers in Washington, Rome, and Paris think that this is the appropriate economic and political policy for Haiti? The Haitian people know Western capitals do not have their interests in mind with their schemes. Most Haitian people reject the Western influence that has killed its leaders, committed multiple coup d'etats, occupied the country many times, and systematically exploited the country.

It is important to note that Aristide, a former Catholic priest, was not a Marxist revolutionary calling for massive confiscation and redistribution of wealth and income. In the words of Aristide, he only wanted what Toussaint L'Overture wanted for Haiti: "freedom, democracy, and dignity." I am not Haitian, but I stand with the Haitian people.

7.11 The Lessons of Uncle Tom's Cabin and Trump's Immigration Plan

November 7, 2024

"I am George Harris. A Mr. Harris, of Kentucky, did call me his property. But now I'm a free man, standing on God's free soil; and my wife and my child I claim as mine. Jim and his mother are here. We have arms to defend ourselves, and we mean to do it. You can come up, if you like; but the first one of you that comes within the range of our bullets is a dead man and the next, and the next; and so on till the last." Uncle Tom's Cabin Harriet Beecher Stowe.

I recently listened to an excellent audio version of this classic 1852 book by Harriet Beecher Stowe that was motivated by the 1850 Fugitive Slave Act that allowed Southern slaveholders the ability to track down runaway slaves, wherever they might be, and bring them back into bondage, regardless of the consequences to the slave and the unity of their families. Ms. Stowe was an outraged abolitionist who knew her book, by highlighting the barbarity and inhumanity of White Southern slaveholders and their Northern industrial collaborators, would demonstrate and lay bare the fundamental and deep contradictions between American slavery, the American Constitution, and Christianity.

There were two reasons for my revisiting this classic work that sold more copies than any book published in the nineteenth century except, ironically the Bible. The first was to re-read and better understand the troubling image of Uncle Tom. Uncle Tom, as imagined, particularly in the minds of twenty-first century Americans is someone that resides deep in the psyche of Black Americans. And secondly to help me imagine what a stated Trump Administration policy of hunting down and rounding up one million "illegal" immigrants could look like based on the very real history of Americans hunting down Africans who sought refuge in free states and Canada. (The Ohio River was known as River Jordan to Blacks who sought to cross from Kentucky and slavery to Ohio and freedom.)

On the first reason, Uncle Tom is probably the most derogatory definition that one Black American can label another Black American. It is often spoken with derision and disgust at an individual who through their actions is a traitor to his race and a denier of his or her Blackness. If this is your view of Uncle Tom, it is a common one, and it is also wrong. And if you do not believe me, you probably have never read the book that bears his name. You probably have based your opinion of Uncle Tom on the portrayal of him by some racist plagiarist who stole the character and completely reversed their view of him in some minstrel show designed to dehumanize and make Blacks feel inferior to Whites. In the mid-nineteenth century, authors had no protection for their work. There were no fewer than a dozen books, plays, shows, and stories that stole Stowe's work and revised it without permission. These bastardized copies were done specifically to justify slavery and its brutality and transform Uncle Tom from a Christ-like hero to a fool.

The real Uncle Tom was a principled Christian man who willingly sacrificed his body, but never his soul to protect his fellow slaves and his family. In my re-reading, I grew to admire Uncle Tom, even though I am a non-Christian. The admiration is not because of what he believed, but because he believed in something and lived his beliefs when most people just talk about what they believe and are quick to throw those beliefs out the window at the first sign of distress. I doubt I would have his courage, or many of you would in the face of the physical brutality he experienced. Read the book.

The second reason for re-reading was because I see in the incoming Administration the same forces that America experienced during slavery about to take place now. Instead of runaway slaves being hunted down and returned to hell on earth, it will be immigrants who will be hunted down and returned to countries we had a hand in destroying. Slavery, notoriously separated mothers from children and husbands from wives. What will happen when one million immigrants are hunted down? Who will do the hunting? Where will they be warehoused? What will be permissible for hunters to do to the immigrants? Will you shelter a fleeing mother and child; one with papers and one without? And what will come of their souls, your souls, and the soul of America?

Stowe's book brought the reality of slavery to American homes. When she visited newly elected President Lincoln, he reportedly said to her, "So you are the little lady that started this great Civil War." Immigration is a problem, but the "solutions" being considered by the incoming Administration will prove more destructive than anything this country has ever experienced since the end of slavery.

"Those who do not learn from history are doomed to repeat it." George Santayana.

7.12 Elon Musk—You Are Fired!

February 21, 2025

I started following Tesla after visiting one of their early showrooms in downtown Washington, DC. At that time, the only thing the showroom had to show was the drive train, the electric motors, and a video of the Model S that was coming out later that year. I was beginning to see the sleek, quiet, electric luxury car on local roads, but it was just a sprinkling not yet a flow.

A few years later I got my first ride in a Model S while visiting Los Angeles. One of my very successful physician friends purchased a Model S. I did not go out to LA simply to ride in his Tesla, but it was the most anticipated part of my trip. After the thrill of instant acceleration, I decided immediately that this was a car, a technology, and a company I wanted to support with my next car purchase.

In late 2018, I bought a performance Model 3. The car was beautiful, powerful, and was a joy to drive. In the first three years of ownership, the car never went to the shop for anything. The only fluid that needs replacing is the window washing fluid. My two Springer Spaniels loved it.

There is a function called Dog Mode that allows you to leave your dogs in the car with a set temperature, even on the hottest days and not worry about their comfort. The front trunk was a useful feature for keeping groceries from melting or keeping hot food hot. It seems Tesla designers had thought of everything,

The only time the car was in the shop was after a deer decided to cross the Merritt Parkway and run into the driver's side of the car. Even though the driver's side glass window shattered into a thousand little harmless pieces. I walked away without a scratch. Unfortunately, the same could not be said for the deer.

Against the advice of my financial adviser, I bought Tesla stock in 2017. I held on to it for a few years. I was a real believer in the company.

I wrote Musk's initial behavior off as the behavior of another eccentric tech-bro genius. After all, only a madman or a genius would want to bore a hole in the earth so his cars could ride unimpeded from California to New York. The car was still the best car I had ever owned, and I have owned many cars in my 70 years.

But when Musk aligned himself with Trump during the 2024 presidential campaign, I had had enough. I would still drive the car, but I was no longer a proud Tesla owner. I stopped wearing my Tesla jacket, which I later donated. I no longer offered friends and associates rides. I was embarrassed to be driving the car from a company owned by a man that seemed to have gone completely off the rails.

After January 20, 2025, and the new administration took over the reigns of power and President Trump placed Musk at the head of the Department to Government Efficiency (DOGE) and the two of them launched their relentless attacks on Diversity, Equity, and Inclusion (DEI) and the federal workforce. Those attacks continue to this day. Their actions sealed the deal for me. The car had to go.

I know selling my car has no impact on Musk's multibillion dollar wealth. He could care less about me or my decision. He has enough wealth to keep the next ten generation of Musks in the millionaire class. I suffer no delusions about my impact on him. I sold the car for my own mental and moral health.

I am happy to report I sold my Tesla 2018 Model 3 and bought a new EV from another company. It feels great getting Musk out of my garage. I actually feel sorry for Tesla cars, in the same way, you feel sorry for an abused dog, who for no fault of their own has a horrible owner. They deserve better. If you are a Tesla owner, you deserve better.

I called my friend in LA, who had upgraded a couple of years back from his Model S to a Model X, to see how he was responding to Musk's actions. He shared that he too was sick of him and had ordered a new electronic SUV from one of the many options that did not exist 3 years ago and was also selling his Tesla Model X in personal protest to Musk's actions. I am reading about other Tesla owners around the world who are doing or considering similar actions. A Canadian Tesla owner set his Tesla on fire in protest. Country music star Cheryl Crow sold her Tesla in protest and donated the proceeds to National Public Radio, one of the organizations targeted by DOGE. I encourage Tesla owners who do not like what Musk is doing to DEI and the federal workforce to sell their Tesla's. I would not advise burning them!

When I was teaching strategy to minority business entrepreneurs at the Tuck School of Business Executive Education program, one of the things we emphasized was the importance and value of brand image. It seems Musk is learning the hard way that your actions and words in business matter to consumers, shareholders, workers, and others. There are already calls for Musk to step aside from the CEO's position at Tesla and let someone else run the company who can focus their entire time and attention on the company. Shareholders are telling Musk his actions are hurting their financial interests.

He probably will not listen to his shareholders. He clearly is not listening to his advisers. But in business, except when you have a monopoly, consumers have the ultimate say. And consumers and shareholders are saying to Musk shut up and do your job as CEO of Tesla or get out. Sales of Tesla are flat and falling in most markets in the United States and Europe. Tesla's market share is already down from over 60 percent to 52 percent in California, its biggest domestic market. Sales of Tesla this year in Germany are down 59 percent!

I do not know what will happen next, but it does not look good for Tesla or Musk. But I am happy to say; "Elon Musk, you are fired!".

CHAPTER 8

Epilogue—Where Do DSP Go from Here?

Abstract In this final chapter, the author discusses the paths Black Americans have taken to accomplish economic and political security. Those paths can be broadly categorized as integration and assimilation or economic and political separation. This discussion of racial strategies for Black Americans dates back to early nineteenth century, and these remain the same paths today.

Keywords Black Americans · Descendants of Stolen People · Segregation · Separation · Reparations · Institutional Racism · Community

"The truth is an offense, but not a sin." Bob Marley from "Jah Live"

This book covers critical issues facing America. The most important and perhaps least addressed of these is centered on wealth. Wealth in America

---

The original version of the chapter has been revised. A correction to this chapter can be found at
https://doi.org/10.1007/978-3-031-93217-5_9

F. McKinney, *Critical Issues Facing America*,
https://doi.org/10.1007/978-3-031-93217-5_8

has become more skewed towards the wealthy than at any time in recent history. According to the Federal Reserve, in 1989, the wealthiest 1 percent of the population owned 22.8 percent of the wealth compared to 31 percent in 2021.[1] It is important to put this in perspective. The top one percent, about 3.5 million Americans, own approximately $44 trillion of the total $140 trillion in household wealth while 167 million Americans in the bottom fifty percent of the population own $3.6 trillion.[2] Wealth inequality is at the root of the political problem facing America with the wealthiest Americans buying politicians, controlling the regulation of industry, destroying the environment, and damaging the general health of the population.

The recent 2024 presidential election showed how destructive wealth inequality is to the American political process. After Harris' announcement to run for the presidency, 40,000 Black women convened on a fundraising call. They were able to raise in that one call $1.4 million. At the time this seemed like a great start for the Harris campaign. Soon after this, Elon Musk announced he would contribute $250 million to the Trump campaign. In exchange for this contribution—even though there is no written agreement the public has seen—Musk was put in charge of the Department of Government Efficiency (DOGE). DOGE in the first few weeks of the Trump administration has taken a wrecking ball to the federal bureaucracy. The billionaires are achieving the first phase of what they want from Trump—a massive tax cut for the wealthy, the elimination of the federal government's countervailing force to monopoly capital and the destruction of federal programs that benefit Black Americans.

The wealth distribution problem is also the most critical problem facing the Descendants of Stolen People. Median Black household wealth was estimated to be $24,500 in 2021 compared to $250,000 in median wealth for White households.[3] The wealth gap impacts all aspects of the

[1] Board of Governors of the Federal Reserve System (US), Share of Net Worth Held by the Top 1% (99th to 100th Wealth Percentiles), July 1, 1989 to April 1, 2025 [WFRBST01134], retrieved from FRED, Federal Reserve Bank of St. Louis; https://fred.stlouisfed.org/series/WFRBST01134, May 1, 2025.

[2] Sullivan, B., Hays, D., & Bennett, N. (2024, April 23). Wealth by race of householder. *U.S. Census Bureau*. https://www.census.gov/library/stories/2024/04/wealth-by-race.html.

[3] U.S. Census Bureau. (2024, April 25). Wealth by race of householder. *U.S. Census Bureau*. https://www.census.gov/library/stories/2024/04/wealth-by-race.html.

quality of life in the DSP community. With negligible household wealth, it is difficult for households to own businesses, housing, and financial assets. It is no surprise given these numbers that only 46 percent of Black households are homeowners compared to 74 percent of White households.[4] White households use their wealth advantage to own, develop, finance, and purchase small businesses that can afford to employ their family members and other members of their community.

Recent efforts in Connecticut creating "Baby Bonds" is a welcome effort to reduce the wealth gap. Baby bonds are bonds provided to low-income Connecticut babies born in the state that mature in 18 to 30 years. Initial investments of $2,500 are put in an account for the child. In 20 years, depending on rates of return, the account can grow to $15,000 to $20,000. This is significantly more than a child born in poverty might have and is welcome resources, but it is unlikely to reduce the racial wealth gap.

The wealth gap also makes it difficult for Black businesses to grow their businesses. Household wealth is one of the key sources of collateral for entrepreneurs seeking business loans and investments. This lack of DSP community wealth inhibits Black business growth and competitiveness.

The wealth gap also promotes political dependency. Black candidates for public office, like the situation Black businesses face, suffer from a lack of capital. Because the DSP community is less wealthy, it does not have the financial resources other communities have to promote their interests in the political marketplace.

The wealth gap also impact the media market. Without large-scale newspapers, cable news networks, and the owners of social media sites, DSP are dependent on getting their information from White owned and controlled media. This lack of substantial media reduces the influence Black leaders have on national political issues impacting the DSP community.

The wealth gap is at the root of all the critical issues addressed in this book. Closing this gap is only possible with a radical new approach. Where do DSP's go from here? Improving the quantity and quality of education, reducing overt housing and labor market discrimination, increasing opportunities for DSP-owned businesses are all things that need to be

[4] Zhao, N. (2024, February 6). Homeownership rates by race and ethnicity. *Eye on Housing*. https://eyeonhousing.org/2024/02/homeownership-rates-by-race-and-ethnicity-3/.

done. However, these traditional solutions will never close the wealth gap between DSP and White Americans. This harsh conclusion is the result of simple math. If you start with the median wealth of $24,500 for DSP households and $250,000 for White households and assume an annual return on wealth of 10 percent for both households, in ten years, DSP median household wealth will equal $66,322 while the median household wealth for White households will grow to $677,000. While the DSP wealth as a percentage of White household wealth remains at 10 percent, the dollar gap has increased from $225,500 to $610,678. The racial wealth gap in America will not be closed by DSP by traditional methods that have been advocated in the past.

Despite this significant and ever widening wealth gap, Black Americans are not without hope or resources. It is estimated that DSP total consumption of goods and services in 2021 totaled over $900 billion. This is more than the Gross Domestic Product of Turkey, Switzerland, Poland, Argentina, and over 200 of the 230 nations of the world. This consumption power has never been used for the widespread benefit of DSP, except when Dr. King and the SCLC brought Montgomery, Alabama to its knees during the 1957 bus boycott.

Before suggesting my solutions, it is important to review some of the history of how DSP leaders have thought about the solutions to DSP problems of racial discrimination and poverty. Understanding DSP strategies to ameliorate these problems provides clues to the solutions recommended here.

8.1 Integration or Separation

Martin Luther King's August 16, 1967, speech entitled "Where Do We Go from Here?" was presented at the tenth annual conference of the Southern Christian Leadership Conference (SCLC) in Atlanta, Georgia. In it he covers the history of the struggle for Black dignity and rights the SCLC led since the Montgomery Boycott of 1957. The strategies outlined in the speech start with the vision presented at the 1963, "I Have a Dream Speech" at the Lincoln Memorial. That dream was a dream of racial integration and full participation in American society. This meant full voting rights, full civil rights, the end of racial segregation in education, healthcare, government service, business participation and other aspects of living. This epilogue asks a similar question Dr King asked almost 60 years later. What will it take for DSP to achieve full equality?

The timing of the "Where Do We Go from Here?" speech came at an inflection point in the struggle for Civil Rights. Other voices had developed since the 1963 "I Have a Dream" speech that presented viable alternatives to integrationist strategies and goals. The Black Power movement represented by luminaries such as Stokley Carmichael, H. Rap Brown, Eldridge Cleaver were putting words into action in northern and western Black inner cities. The Black Panther Party organized daycare centers, free breakfast, and lunch programs, Black education programs, self-defense and gun safety classes, and political action and organizing efforts. Up until the assassination of Malcolm X in 1965, the Black Muslims also provided a viable alternative to the integrationist perspective. Thousands of Blacks in northern and western cities dumped their Christian upbringing and became Black Muslims because of their messages of clean living and Black self-determination. And a combination of socialists and communists were on the margins seeking inroads into the Black community, but the efforts by socialists and communists were largely unsuccessful.

Today, these two strategic paths; Integration and Separation, remain the primary choices for DSP. The Integrationist approach of full inclusion into American society is explicit in the DEI goals of post George Floyd America. Inclusion in business, government, the arts, housing, and the nonprofit world is consistent with the conservative meritocratic vision of the American pyramid. Much of my professional career as a proponent of supplier diversity is consistent with this integrationist vision. In the case of supplier diversity, the goal was the inclusion of "diverse" firms in corporate and public sector supply chains. The integrationist vision is also consistent with my membership in the oldest Black fraternity, Sigma Pi Phi (the Boule), that caters to what Dubois called the talented tenth. (Dubois was an early member of the Boule.) The Links served a similar role for Black women who saw themselves as beacons and exemplars for Black women lower in the American pyramid. These and other elite Black organizations not only sought camaraderie but also sought to leverage their educational and cultural advantages into higher paying, higher status positions within the White dominated American pyramid.

The largely internal debate within the Descendants of Stolen People community did not start with Dr. King v. Malcolm X in the 1960s. DSP from their first days in captivity struggled with the nearly unbearable weight of their conditions. Some of those first captured Africans decided to jump off the slave ships into certain death, rather than continue their journey over the middle passage. Africans continued to struggle in small- and large-scale slave revolts throughout the eighteenth and nineteenth centuries. Slaves ran away from their captures' plantations seeking some semblance of freedom. Slaves that did not runaway or kill their enslavers, sabotaged the production on plantations across the South. This was one of the primary reasons contemporary economists during Slavery noted the inefficiency of slavery compared to free labor.

Abolitionists like William Lloyd Garrison had a goal of ending slavery. The first debate within the DSP community before the Civil War was over what to do and where to go at the end of slavery. Groups of "free" DSP met in Philadelphia beginning in 1832 and continuing through 1855 where these questions were discussed. These meetings were comprised primarily of well-educated and relatively well-off DSP. The records of these meetings provide a preview of future debates. These early DSP strategists contemplated what should follow slavery. Their solutions ranged from going back to Africa, full integration, and full separation within the boundaries of the United State. Frederick Douglas was one of the primary advocates for full integration here within the United States. The return to Africa supporters outside the DSP community had high-profile support, initially from such luminaries as Abraham Lincoln and the KKK. Liberia on the West African coast was established by the American Colonization Society in 1822 with Liberia declaring independence in 1847. Thousands of former slaves and free DSP moved to Liberia during the nineteenth century.

At the end of the nineteenth and the early twentieth centuries, another debate over DSP strategy emerged. This time, the primary proponents of the different strategies were W.E.B. Dubois who represented the Integrationist approach, Marcus Garvey who represented the repatriation to Africa approach, and Booker T. Washington who called for full economic integration with social segregation.

While the first DSP debate prior to the end of the slave regime was largely in response to slavery, the late nineteenth and twentieth centuries debate took place in the context of Jim Crow, and the mid-to-late twentieth century debate occurred in the post 1954 Brown v. Board of Education case.

Now we are in a different political and global economic environment combined with a resurgence of fascist movements around the world, led by America under the leadership of Donald Trump. This different context calls for DSP to reopen this two-hundred-year-old debate and seek new answers for the new conditions Descendants of Stolen People face. My suggestions are preliminary but might form the basis of a public airing of the options that will address the needs of the majority of DSP.

8.2 A Path Forward—Black Gentrification and Community Reparations

Knowing White friends and enemies for over 70 years, I know that the solutions I am about to lay out may offend some of them. It will offend my White friends who hope for an integrated American society, and it will offend my White enemies because it tells the truth about them and their desire to never allow for the full development of the Descendants of Stolen People. These views will also offend many within the DSP community because they hold on to the fantasy that continuing to do the same things will lead to different outcomes when it comes to Black wealth and community development. One of the primary responsibilities of a social scientist is to describe reality regardless of who agrees with or supports that description. "The truth is an offense but not a sin." B. Marley.

The Descendants of Stolen People have tried integrating and assimilating into the American pyramid. There have been notable successes in this attempt at integration and assimilation. We can point to DSP executives who have risen to the ranks of CEO and Board Chair for multiple Fortune 500 companies. We can now point to the election of Barack Obama as providing a North Star showing the possibility of excelling in the nation's best institutions of higher learning. There are now a few DSP billionaires and nine figure millionaires within the DSP community. All these examples are the myth building and supporting examples that keep those at the bottom supporting a pyramidal system. The thinking is if they can do it, I might be able to. There must be another path forward for the DSP. Fortunately, we do not have to create that path in completely unexplored territory. Our ancestors have shown us the way forward.

Marcus Garvey, Booker T. Washington, Black Wall Street, W.E.B. Dubois, Malcolm X, Martin Luther King have each provided aspects of the solution to the problems of DSP in America. But their paths led to different visions and required different strategies. The solution I propose is a synthesis of what can be taken from these different strategies of integration into the American pyramid and separation from the dominant and malevolent forces of White institutions.

Descendants of Stolen People can become as wealthy as any ethnic group in America today if two things happen: Black Gentrification and Community Reparations. If these can happen, DSP can achieve the economic independence our ancestors have dreamed of, and we can free ourselves of dependence on White institutions whose allegiance is to White Americans not the DSP.

The first condition is that higher income and wealthy DSP must physically live within a DSP community. The traditional Jeffersonian solution—I am talking George and Louise aka "Weesy," not Thomas—is to "move on up to the East Side" out of Harlem and into neighborhoods where Whites are the dominate residents. Every American city has areas that were color identified. Redlining of neighborhoods where Blacks were restricted to live was only one of the colors used in this institutionalized segregation of American neighborhoods. There were also blue lines and green lines and yellow lines that provided support for the American pyramid. Blue neighborhoods also known as Type B were still considered desirable neighborhoods for White U.S. born residents. Type C or Yellow neighborhoods were declining neighborhoods and were for White immigrants and were subject to an infusion of DSP residents moving from Redlined or D neighborhoods. Type A neighborhoods or Green neighborhoods were the most "desirable" and "safe" neighborhoods reserved for Whites at the top of the pyramid. The history of institutionalized housing segregation and the American pyramid is well-documented in the excellent book by Richard Rothstein's classic, The Color of Law: A Forgotten History of How Our Government Segregated America.

Beginning in the late 1980s, mostly young, middle-class U.S. born White Americans who were priced out of moving into Type B neighborhoods and certainly Type A neighborhoods began moving into Type C neighborhoods. In 1970, Washington, DC, had a Black population of 538,000 representing 71.1 percent of the total population. By 1990, this was down to 65.8 percent and less than 400,000 DSP residents. This trend continues today where in 2023 it is estimated that the DSP population is down to 297,000 representing 44 percent of the population. DC was once known by its residents as the "Chocolate City" now at best it is a double-stuff Oreo cookie—with whites in the center and DSP in the fringes. This is White gentrification.

The solution for DSP is to use this pattern of resegregation of the nation's largest cities to our advantage. White folks may gentrify a neighborhood that temporarily looks like the integrationist's nirvana, but their goal is for the remaining lower-income Black folks to move out and be replaced by Whites.

Black gentrification should come in two forms: One would be higher income DSP moving back into the same communities White gentrifiers have targeted and make these formerly low-income Black communities economically diverse but racially homogeneous, i.e., to borrow a phrase from the MAGA crowd—Make Black Communities Great Again! This amounts to competition for some of the most desirable locations within major metropolitan areas and will be an expensive proposition. But for some DSP, who understandably want to live in core urban settings, this is one aspect of Black gentrification.[5]

The second form of DSP gentrification would be for DSP to move out of the more expensive urban areas and take over the outer rings of major metropolitan centers where Whites and other racial groups now also have a significant presence. Fortunately, there are nascent examples of this in communities outside of Washington, DC. Bowie, Maryland in Prince Georges County has a town of 57,000 people and 57 percent of the residents are DSP. The median household income in Bowie is $141,000. The median household income in the United States is $79,000, 56 percent of the median household income in Bowie. Bowie also has substantial income diversity within this majority DSP town. In this town, and county, it is no surprise that DSP controls political power and has a substantial number of DSP-owned enterprises. The former Prince George's County Commissioner is now the U.S. Senator from Maryland. The Governor of Maryland is a DSP. And the mayors of Baltimore and Washington, DC, are DSP. This is the model DSP should follow throughout the country.

Black gentrification in these areas should accelerate and be encouraged. These high percentage and high-income DSP communities should continue to encourage DSP with wealth and income to move into these communities, while welcoming middle class and lower-income DSP residents to move in as well.

[5] A third possible concentration of DSP would be DSP living in the rural Southern states. Their concerns are different from urban and suburban DSP, and require a separate analysis that is not included in this book. I have some thoughts on their plight and addressed some of those in a paper on why retired Blacks should move back to states like Mississippi, South Carolina, North Carolina, Georgia, Texas, and Arkansas where the cost of living is much lower and it is possible for DSP to convert towns, counties and even states into majority DSP areas.

Economic diversity within DSP majority communities is what is necessary for DSP to develop the economy that will support the residents in these communities. As these communities grow, so will the number, the quality, and the size of DSP small businesses to support the residents. Black consumers will support DSP businesses.

DSP-owned businesses can develop more easily and effectively if those businesses have a concentrated population of DSP with disposable income. Right now, the concentration of DSP is in low-income communities that do not have the resources to support DSP-owned businesses. But in economically diverse (gentrified) DSP communities, small and midsized businesses can grow and replace White owned and other minority owned businesses that are serving the community today. The responsibility to make this happen rests firmly with the Black elite who must physically move into DSP communities and out of communities where they are the distinct minority. The incomes of the Black elite currently go to support the White owned businesses whose owners are their neighbors.

In Booker T. Washington's famous speech at the 1896 Atlanta Exposition, he said:

> In all things that are purely social we can be as separate as the fingers, yet one as the hand in all things essential to mutual progress.

The solution of Black gentrification in urban areas, and the demographic takeover of outer ring towns turns Booker T. Washington's solution inside out. Instead of being separate on social things and united on economics, Black gentrification calls for the separation of residential communities and economic progress while allowing for social integration between the races. There is nothing in what I propose that requires or calls for animosity between the races. But what it does, unequivocally requires that DSP form their own local economic foundation, separate from White business and professional interests. Only if these conditions are met, will DSP be able to develop the small- and medium-sized businesses that can support their economic, political, and social development.

This also does not call for DSP to stop working for White owned and operated companies, or attending White institutions of higher learning, or being friends with members of the White community.

Black gentrification also creates the conditions for the development of independent political institutions that work in the interests of DSP communities instead of the institutions that now work primarily in the interests of White Americans. For example, majority DSP and economically diverse communities will elect DSP officials. They will change the public education system that will promote DSP culture and history. They will intervene when DSP engage in criminal activities in ways that are reformative instead of simply warehousing them. This abusive system has developed specifically to control DSP. Economically integrated DSP communities will have role models of Black excellence that will inspire DSP youth and families to succeed. These communities will build lasting wealth that will put DSP on a path towards closing the wealth gap in America.

This might be a great idea, but the challenge is in how to make it a reality. Black gentrification requires a level of racial and cultural solidarity that we have not seen since the Marcus Garvey Universal Negro Improvement Association (UNIA). Fortunately, the current Trump administration is having the effect of communicating to even the most conservative DSP that a significant portion of White America is not interested in their economically well-being. This will make collaboration and organizing within DSP communities easier and more relevant to DSP from all levels of income and wealth. From the Boule and Links to lowest income Blacks, the time is now to come together.

8.3 Community Reparations and DEI

It is time for DSP and White America to reach a grand bargain that eliminates the tension between the races that started with the illegal and immoral stealing of Africans from their African homelands and enslaving them here in America. The details of this grand bargain are outlined here and are based on the historic realities of the DSP.

Every time in our history that White Americans are threatened with economic and political insecurity, they attack and scapegoat Black Americans. It happened before the Civil War when Northern White Americans attacked free and enslaved Blacks in Northern cities because they did not want to risk their lives for the freedom of Black Americans. For an excellent representation of this watch the movie, the Gangs of New York where White New Yorkers attacked Black New Yorkers because of the Civil War draft.

It happened again because of Reconstruction when Southern Whites reacted with terrorism, the KKK, and the imposition of Jim Crow laws. It happened after WW I when Black progress was being made in cities like Tulsa Oklahoma where Black Wall Street was burned to the ground in 1921. It happened again during the Great Depression when White Americans created redlined areas to reserve the most desirable locations for Whites only. It happened in the 1970s and 80 s when White Americans attacked affirmative action with dozens of lawsuits claiming reverse discrimination. And it is happening now with the MAGA Trump Administration doing everything in its power to eviscerate Diversity, Equity, and Inclusion from American public and private policy.

The message to DSP is simple: White America and its institutions cannot be trusted to be long-term allies and that it is up to DSP to take care of ourselves.

The grand bargain will have two dominant components; something for White America and its institutions and something for DSP.

For White America and its institutions, DSP would permanently reject all forms of future DEI and affirmative action by the federal government and institutions supported by the federal government. This would mean an end to any and all programs or laws that identify beneficiaries based on race. Gone would be affirmative action in education, the military, government agencies, the courts, the Executive Branch, the Legislative Branch and the nonprofit world. Gone would be any requirement to hire, employ, contract, or do anything with DSP from here forward. We do not need or want their help if they do the following.

In exchange for foregoing all efforts at DEI, the federal government would transfer $3 trillion into a DSP Community Development fund controlled by the DSP community.[6] These are not individual reparations but amount to what be called Community Reparations.

One of the many problems with individual reparations is the lack of good data on DSP. While many DSP can trace their forefathers and foremothers to specific slave plantations, that is the exception not the rule. The other problem with individual reparations is that a significant amount of those funds would be squandered and or stolen. Community reparations is the solution to both of these problems. By distributing

[6] There are over 43 million Blacks in America. Eighteen percent of Blacks are foreign born. After factoring into Blacks who have both sets of grandparents born in the United States, would leave about 30 million Blacks. The $3 trillion estimate works out to about $100,000 per DSP with direct ties to American apartheid. Blacks in the United States from Jamaica, Haiti, Nigeria, Ghana, and other countries need to be paid reparations from English and the French.

the $3 trillion to DSP communities the money would be designated for educational enhancement, housing, healthcare, and entrepreneurial development. These funds could be used to develop DSP-owned enterprises that would primarily serve the DSP community.

White Americans would no longer have the cudgel of DEI as an excuse for their own underperformance and mediocrity. DSP development would also allow for DSP communities within the same state and in other states to connect with one another in ways that promote further economic and political development. These stronger more independent DSP communities could also connect to the wider African diaspora in creative and mutually supportive ways.

Questions of how the $3 trillion would be distributed and how DSP communities are defined are not new problems and they are not unsolvable. These funds would encourage and support DSP gentrification in historical DSP communities and in new DSP communities.

Black gentrification and Community Reparations will be criticized by people within the DSP community and outside of the DSP community, but I think two hundred plus years of failing to integrate and assimilate into the American pyramid is long enough to conclude that what has been tried in the past does not work and will not work for the majority of DSP. It is time to move on.

8.3.1 A Final Note to the White Allies of DSP

The recommendations put forward in this book should not be interpreted as expressing animus to White Americans in general. DSP and many White Americans share similar goals of taking care of their families, living in safe communities, having clean water and air, and harmonious relationships with neighbors domestically and abroad.

The conclusions presented here should also not be interpreted as a desire for DSP to have a separate country from America. A multi-racial democratic society is the North Star for both DSP and White Americans. We should pursue that goal jointly and with respect and love.

Correction to: Epilogue—Where Do DSP Go from Here?

Correction to:
Chapter 8 in: F. McKinney, *Critical Issues Facing America*, https://doi.org/10.1007/978-3-031-93217-5_8

Footnote in chapter 8 was inadvertently published with an error which has now been corrected.

The updated version of this chapter can be found at
https://doi.org/10.1007/978-3-031-93217-5_8

F. McKinney, *Critical Issues Facing America*,
https://doi.org/10.1007/978-3-031-93217-5_9

A Partial Bibliography of Books That Informed This Book and the Author

Baldwin, James, I *Am Not Your Negro*, Vintage International Press, 2011
Baldwin, James, *The Fire Next Time*, Vintage 1992
Becker, The Economics of Discrimination, University of Chicago Press, 1994
Blight, David W., *Frederick Douglass: Prophet of Freedom*, Simon and Schuster, 2020
Carmichael, Stokely and Hamilton, Charles, *Black Power: The Politics of Liberation* Vintage Press 1967
Chernow, Ron, *The House of Morgan*, Grove Press, 2010
Coates, Ta-Nahisi, *Between the World and Me*, Random House 2017
Douglas, Frederick, The Life and Times of Frederick Douglass, 1881
Dubois, W.E.B., *The Souls of Black Folk*, A.C. McClurg, 1903
Dubois, W.E.B., *Black Reconstruction in Ameica:1860–1880*, Free Press, 1992
Ferrer, Ada, *Cuba: An American History*, Scribner, 2022
Foner, Eric, *Reconstruction, 1863–1877: America's Unfinished Revolution*, Harper and Row, 1988
Frazier, E. Franklin, *The Negro in the United States*, Macmillian, 1957
Friedman, Milton and Rose Friedman, *Capitalism and Freedom*, University of Chicago Press, 2002
Galbraith, John Kenneth, *The New Industrial State*, 1967
Garvey, Marcus, *The Philosophy and Opinions of Marcus Garvey, or Africa for the Africans*, Majority Press 1986
Garvey, Marcus, *Selected Writings and Speeches*, Grove Press, 2004
Gugges, David, *The Haitian Revolution*, Hackett Press, 2014
Haley, Alex, *The Autobiography of Malcolm X: As Told by Alex Haley*, 1965

F. McKinney, *Critical Issues Facing America*,
https://doi.org/10.1007/978-3-031-93217-5

Heilbroner, Robert, *The Worldly Philosophers: The Lives, Times and Ideas of the Great Economic Thinkers,* 1953

Keynes, John Maynard, *The General Theory of Employment, Interest and Money*, 1936

King, Martin Luther, *Where Do We Go from Here: Chaos or Community*, 1967

King, Martin Luther, *Why We Can't Wait*, 1963

Kinzer, Steven, *The Brothers: John Foster Dulles, Allen Dulles, and Their Secret World War: John Foster Dulles, Allen Dulles, and Their Secret World War*, Times Books, 2013

Lewis, David Levering, *W.E.B. Dubois: Biography of a Race1868–1919,* Henry Holt and Company 1994

McKinney, Rufus, William, Sr. *Beating the Odds: The Story of One Black Man's Life in 20th Century America*, Vantage Press, 2005

Myrdal, Gunnar, *An American Dilemma: The Negro Problem and Modern Democracy*, Pantheon Books, 1975

Piketty, Thomas, *The Economics of Inequality,* Belknap Press, 2015

Rodney, Walter, *How Europe Underdeveloped Africa,* Verso, 2018

Samuelson, Paul and William Nordhaus, *Economics*, McGraw-Hill, 2009

Smith, Adam, *An Inquiry into the Nature and Causes of the Wealth of Nations,* 1776

Sowell, Thomas, The Thomas Sowell Reader, Basic Books 2011

Stiglitz, Joseph, *The Price of Inequality: How Today's Divided Society Endangers our Future*, WW Norton, 2012

Stowe, Harriet Beecher, *Uncle Tom's Cabin*, 1852

The most frequently viewed site for economic data was the Federal Reserve Economic Database (FRED) published and managed by the Federal Reserve of St. Louis.

Washington, Booker T., *Up From Slavery: An Autobiography*, Doubleday, 1963

Wilkerson, Isabel, *Caste: The Origins of our Discontents*, Thrift Books, 2002

Wilkerson Isabel, *The Warmth of Other Suns*, Vintage Press, 2011

Woodson Carter Goodwin, *The Mis-Education of the Negro*, 1933

Woodward, C. Vann, *The Strange Career of Jim Crow,* Oxford University Press, 1974 (1955 originally published)

X, Malcolm, *Malcolm X Speaks: Selected Speeches and Statements, Introduction by Ibram Kendi* Grove Press, 2024

MIX
Papier aus verantwortungsvollen Quellen
Paper from responsible sources
FSC® C105338

If you have any concerns about our products,
you can contact us on
ProductSafety@springernature.com

In case Publisher is established outside the EU,
the EU authorized representative is:
Springer Nature Customer Service Center GmbH
Europaplatz 3, 69115 Heidelberg, Germany

Printed by Libri Plureos GmbH
in Hamburg, Germany